Contents

Content Guidance

Questions & Answers

Getting the most from this book

Questions & Answers

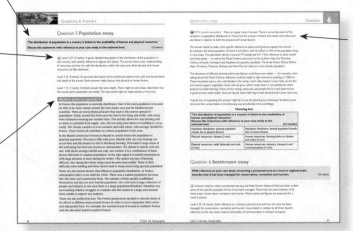

About this book

Much of the knowledge and understanding needed for AS geography builds on what you have learned for GCSE geography, but with an added focus on geographical skills and techniques, and concepts. This guide offers advice for the effective revision of **Unit AS 2: Human Geography** (including skills and techniques), which all students need to complete.

The AS 2 external exam paper tests your knowledge and application of geographical skills and techniques and lasts 1½ hours. The unit makes up 50% of the AS award or 25% of the final A-level grade.

To be successful in this unit you have to understand:
- the key ideas of the content
- the nature of the assessment material — by reviewing and practising sample structured questions
- how to achieve a high level of performance within the examination

This guide has two sections:

Content guidance — this summarises some of the key information that you need to know to be able to answer the examination questions with a high degree of accuracy and depth. In particular, the meaning of keys terms is made clear and some attention is paid to providing details of case study material to help to meet the spatial context requirement within the specification.

Questions and answers — this includes some sample questions similar in style to those you might expect in the exam. There are some sample student responses to these questions as well as detailed analysis, which will give further guidance in relation to what exam markers are looking for to award top marks.

The best way to use this book is to read through the relevant topic area first before practising the questions. Only refer to the answers and examiner comments after you have attempted the questions.

Content Guidance

Skills and techniques

The first question on the AS Unit 2 examination paper will test your knowledge and application of a series of geographical skills and techniques — from either the physical or human side of the course. You will be expected to respond to a range of quantitative and qualitative questions. The diversity of data collection and processing opportunities and stimulus material in A-level geography means that it is impossible in this short guide to refer to every possible map or graph type available.

You are expected to have knowledge of:
- **data collection:**
  - using surveys and questionnaires
  - analysis and interpretation of maps at a range of scales, photographs, remotely sensed images and data from secondary sources (see map skills below)
- **data processing:**
  - map skills — map distributions, densities and flows using dots, flow lines, choropleth shading and isolines
  - sketches — drawing annotated sketch maps
  - graphical skills — constructing, analysing and interpreting scatter graphs, line graphs, bar graphs, pie charts, proportional graphs and triangular graphs (including titles, keys, scales, frames and direction arrows)
  - sampling methods — sampling, including random, systematic, stratified (point, line and quadrat) and pragmatic
  - statistical analysis — using mean, median, mode and range; Spearman's rank correlation and nearest neighbour analysis
- **embedded skills** — although not specified in detail, you are expected to be able to use, and refer to use of, geographic information systems (GIS) and the internet. You are encouraged to use ICT for collecting, sorting, recording and presenting geographical information. These skills are not covered in this guide.

Data collection

Much of the information on fieldwork/data collection can be found in the AS Unit 1 Student Unit Guide. However, some of the data collection methods are focused on human geography and are more likely to come up in Unit 2.

Using surveys and questionnaires

Surveys and questionnaires are used in many different ways in human geography. Data comprise the information that you collect in order to address the aims or hypotheses within any investigation. Questionnaires can be a good method of collecting information and opinions that people have. You need to think carefully about the sample size (what is the appropriate number of questionnaires to ask in relation to a study?) and also consider the number of questions and length of the questionnaire — it always takes longer than you expect to administer a questionnaire.

- Primary sources/data refer to new information that you have collected in the field. This might be done through observation or through measurement.
- Secondary sources/data refer to information that has been obtained from any other source.
- You will often need to use both primary and secondary information.

The main sampling techniques are random, systematic, stratified (point, line and quadrat) and pragmatic. Sampling techniques are needed to reduce bias in the investigation. Some of the techniques introduce more bias into the study than others. (See below for more information on sampling methods.)

Data processing

Map skills

Photos

Questions might ask you to label or annotate a photograph such as Figure 1. Each label should be a word or sentence that describes or identifies a feature shown in the photograph.

Michael Raw

Figure 1 A river channel

Satellite images

Satellite images might be used to allow students to analyse and interpret geographical patterns within either a physical or human environment. You need to use each resource carefully and should look for patterns on the map and describe these geographically.

Dot distribution maps

Dot maps can be used to show a distribution pattern within an area. Each dot will represent a specific value and the number of dots in an area will indicate the density and distribution of this within the population.

Figure 13 on page 20 is an example of a dot distribution map. It shows the distribution of Met Office automatic weather stations across Northern Ireland.

Flow line maps

Flow lines can be used on a map to show the amount/volume and the direction of a movement from one place to another (Figure 2). Usually, the width of the flow line will indicate the amount of movement in this direction (used in conjunction with a scale key). These maps can be useful to illustrate, for example, migration streams or flows of tourists from one place to another.

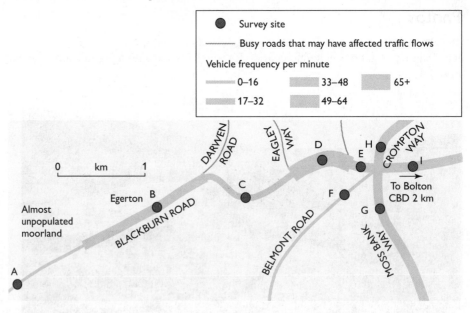

Figure 2 Traffic flows on the main roads to the north of Bolton, Greater Manchester

Knowledge check 2

(a) On which section of road is the highest amount of traffic?

(b) Name two roads where there is very little traffic.

Choropleth maps

Choropleth maps (sometimes called 'area-shaded' maps) are shaded according to the density of a value in an area (Figure 3). Usually, any data used in the construction of a choropleth will be grouped into around 5–7 different categories. These will then be plotted and shaded appropriately on the map.

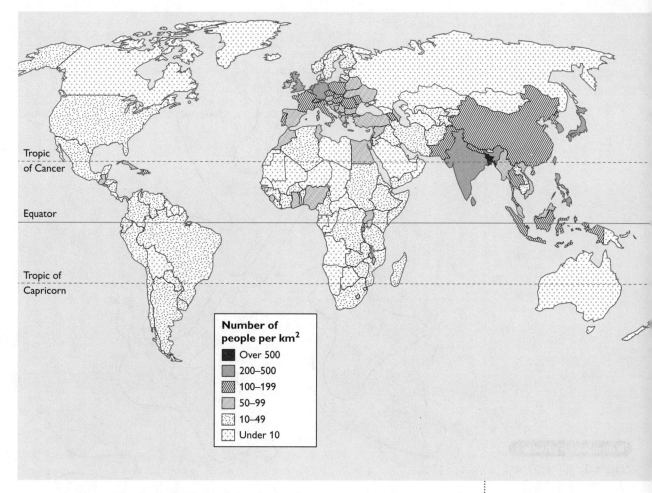

Figure 3 World population density by country, 1996

Some geographers argue that these maps oversimplify the data because they can suggest that a large area has the same value. For example, in Figure 3, all of China is shown as having between 100 and 199 people per km². However, this does not reflect the fact that some parts of China are much more densely populated and some are sparsely populated. Maps like this often suggest that there are huge changes at boundaries (e.g. the UK and Ireland) whereas this might not be the case.

Isoline maps

An isoline is a line that joins places on a map that have the same value (Figure 4). Some of the most commonly used isolines are isobars (where areas of the same pressure on a weather map are joined), isotherms (where areas of the same temperature are joined) and contours (where areas with the same height on an OS map are joined).

Examiner tip
Choropleth maps are common in exam papers as they can challenge the student to think carefully. Think about how you might go about constructing a choropleth map. What are the advantages of displaying your data in a format like this?

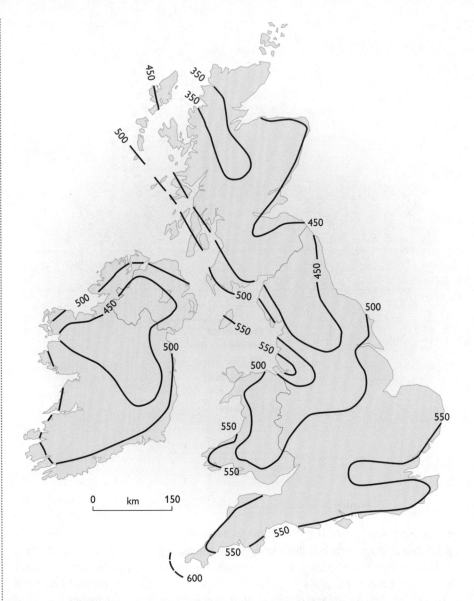

Figure 4 Potential evapotranspiration in the British Isles (mm)

Knowledge check 3

(a) Add a 400 mm line to Figure 4.

(b) Describe the pattern of potential evapotranspiration in the UK (don't forget to describe the location and use figures to support your answer).

Sketches

Field sketches

These are similar to the annotated photos mentioned above. The key is to use labels and annotations to describe and explain some of the features and processes shown in each resource.

Annotated sketch map

Sometimes, examiners might ask students to draw a sketch map from a grid square or series of grid squares on a map. You are usually given an OS map and asked to copy some of the main features onto an enlarged, blank grid square. When drawing a sketch map remember to:

- add a title
- use as many labels as possible to describe what you see on the map
- add a scale
- add a key to show any features you have added
- show any contours
- add a north arrow
- mark any grid squares

Graphical skills

Bar graphs

Bar graphs (Figure 5) are common in geography and can be presented using a number of different formats.

A common misconception is that bar graphs and histograms are the same. A histogram is usually used to display continuous data and has blocks connected to each other. It also shows numbers (or frequencies) of a whole sample and should be all shaded in the same colour.

Bar graphs are usually used to display discrete (non-continuous) data, with the blocks separated by a space. As each bar represents different, discrete data, different colours can be used for each bar.

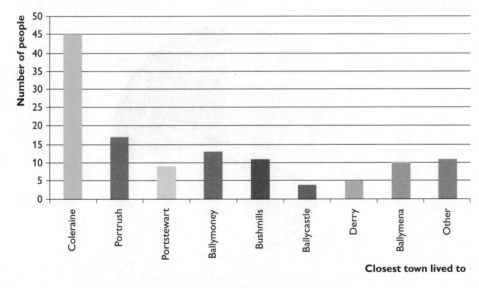

Figure 5 A bar graph to show the town nearest to where people live in a population survey in Coleraine town centre

Examiner tip

Always draw maps and graphs neatly and accurately so that you maximise your mark in the exam.

Knowledge check 4

Is Figure 5 a bar chart or a histogram? Analyse (describe) and give a geographical explanation (interpretation) of the pattern in Figure 5.

Examiner tip

Make sure you know the difference between a bar graph and a histogram.

Line graphs

Line graphs (Figure 6) can be effective for displaying continuous data or measurements and can be used to show changes over time.

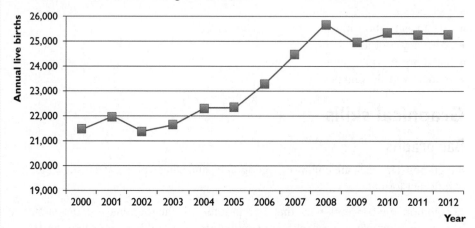

Figure 6 A line graph showing live births in Northern Ireland, 2000–2012

Proportional graphs

Pie charts are a good method of displaying data proportionality. A pie chart is divided into segments with angles proportional to the data (Figure 7). The pie circle represents 100% of the data set.

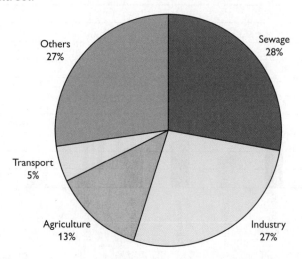

Note: 'Others' are incidents in which the source cannot be accurately identified.

Figure 7 River pollution incidents in England and Wales, 1990

Sometimes the size of each pie chart is used to indicate the size of a population, and this can be compared to a scale.

Scattergraphs

Scattergraphs are slightly more complicated because they often involve two different types of observation. Each observation is called a variable. Scattergraphs plot the relationship or correlation between the two variables (Figure 8).

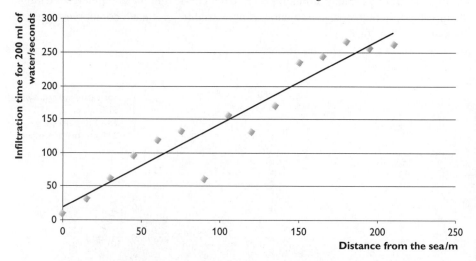

Figure 8 Scattergraph showing infiltration rate through a sand dune against distance from the sea

Once the graph is drawn the strength of the relationship can be tested using a 'line of best fit'. The graph might then look like one of four possibilities (Figure 9).

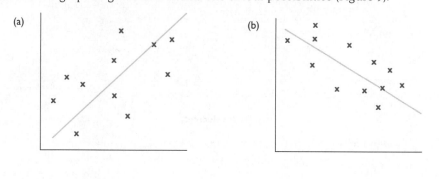

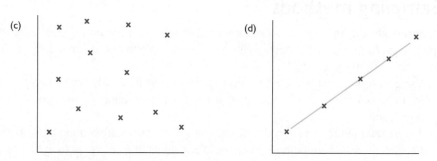

Figure 9 Scattergraphs showing (a) strong positive correlation, (b) strong negative correlation, (c) no correlation, (d) perfect positive correlation

Examiner tip
Scattergraphs can take time to perfect, so make sure that you practise drawing them.

The drawing of a scattergraph is usually the starting point for the Spearman's rank correlation statistic.

Triangular graphs

Triangular graphs are plotted using three axes connected in an equilateral triangle (Figure 10). It is only possible to show three variables and each component must be measured out of 100%.

Knowledge check 5

(a) What is the breakdown of France's total electricity production?

(b) What is the breakdown of Taiwan's total electricity production?

(c) Plot the results for New Zealand onto the graph: 72% thermal and other, 0% nuclear, 28% hydroelectric.

Examiner tip

Many students struggle to understand triangular graphs — make sure that you know how to use them.

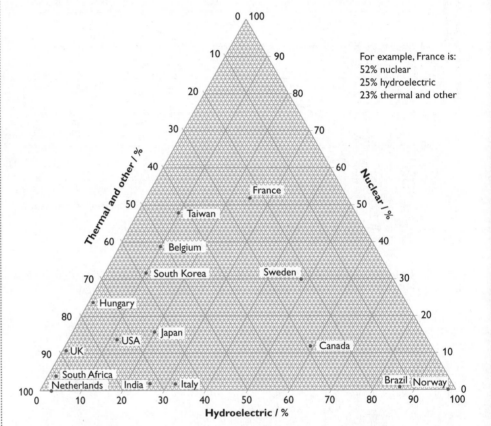

Figure 10 Triangular graph showing percentage of total electricity production by generating source for selected countries (1988)

Sampling methods

Random sampling is when a random number table or random number generator app is used to give the sequence of, for example, people to ask or houses to call in a street when doing a survey.

+ Using a random number table means that results will be totally random and should be unbiased. This should also give a good representation of the whole area/population.

– The technique takes no account of any changes/subsets/differences in an area or population, so a population survey could end up asking only people aged over 65 (this is called 'bunching').

Systematic sampling is when samples are taken using a pre-determined interval. Questionnaires might target every 5th or 10th person who walks past, soil studies might be taken every 5 m or 10 m along a survey line/transect.

+ This is best for studies that are measured over time or over a particular distance. It is suitable for studies like sand dune surveys or measurements taken across a river.
– There is an increased chance of bias as it is the individual researcher who decides on the interval. Fixed intervals might mean that important measures (like the slope angle on a sand dune) might be meaningless, as the upward or downward slopes might not appear in the results.

Stratified sampling is useful when there are clear sub-groups in the dataset. For example, if conducting a sample of 70 questionnaires in a school population of 700 students, you might break the sample down so that equal numbers of questionnaires are given to each year group — 10 to year 8, 10 to year 9, 10 to year 10 etc. You could break this down even further and ask for two questionnaires from each of five form classes and again break things down to ask one boy and one girl from each form class.

+ This can give a broad and detailed snapshot of what a population/group thinks about a particular issue.
– This can be a complicated and difficult method of collecting results, taking a lot of time and organisation.

Point sampling is when individual points are used in the investigation. For example, specific accessible sites might be chosen in a river study.

Line or **transect sampling** is when a line is drawn on a map in an area and all data collected along this line.

Quadrat sampling is when a piece of equipment called a quadrat is used to measure the amount of vegetation/type of vegetation or amount of ground coverage within an area. The quadrat is usually a square metal frame (the most common is 50 cm × 50 cm), which is placed onto the ground during an ecological study.

Pragmatic sampling is when decisions are taken to visit sites that are safe/accessible or which might demonstrate typical characteristics. Although this approach often allows for a simple fieldwork experience, it introduces a huge amount of bias into the sampling technique.

Examiner tip

Make sure that you know the positives and negatives of using each of the different sampling methods. Each one will be more useful for a particular type of investigation.

Statistical analysis

Many students struggle with the statistical analysis section of AS geography but these different statistical techniques are relatively straightforward to use and understand.

Mean, median, mode and range (measures of central tendency)

The **mean**, often known as the average, is found by adding together all the values under investigation and dividing this by the total number of values.

+ It can be an accurate measure as all values are seen as equally important, and it is a relatively simple calculation.
– It can be distorted if there is one extreme value, and sometimes the decimal places can cause confusion. For example, if a total fertility rate is 2.3, how can the average mother have 0.3 of a child?

Knowledge check 6

Each student in a class of 19 students noted the age of their eldest surviving grandparent:

90, 89, 80, 76, 65, 54, 85, 75, 81, 76, 73, 72, 66, 74, 69, 70, 64, 62, 77

(a) What are the mean, median and mode for these data?

(b) Which of these measures do you think is the most useful, and why?

(c) Which of these measures do you think is the least useful, and why?

The **median** is the central value when the values are ranked in a series. If there is an even number, the median is the mid point between the two middle values.

+ It is a good way of finding the 'centre' of the distribution pattern within the data set and is not usually affected by extreme values.

– It is less reliable when there are few values and it cannot really be used for any further mathematical processing.

The **mode** is the most frequently occurring number in the data set.

+ It is quick to calculate and can help to describe the general distribution of the data.

– It has limited value as the modal value could end up being one of the extreme values, which would cause bias in the result set. It does not take into full consideration the full range of values in the data.

Spearman's rank correlation

Drawing a scattergraph is often the first step in trying to understand the relationship between two variables or two groups of data in a fieldwork study.

The Spearman's rank correlation coefficient (r_s) is a statistical test that shows the strength of the correlation or relationship between the two groups of data. It shows the strength and type of relationship.

Usually, the Spearman's rank correlation only works when there are at least 15 sets of values. The first step is to think of a hypothesis to test. In this example we investigate whether 'there is a negative correlation between global birth rates and GNI per capita'.

Table 1 Spearman's rank correlation for birth rate against GNI

Country	Births per 1000 population	Rank	GNI PPP per capita (US$) 2010	Rank	d	d²
Burkina Faso	43	2	1,250	15	−13	169
Kenya	35	3	1,640	13	−10	100
Zambia	46	1	1,380	14	−13	169
Tunisia	19	7.5	9,060	10	−2.5	6.25
South Africa	21	5	10,360	9	−4	16
Canada	11	14	38,370	1	13	169
Mexico	20	6	14,400	7	−1	1
Jamaica	16	9.5	7,310	11	−1.5	2.25
Argentina	19	7.5	15,570	6	1.5	2.25
Brazil	16	9.5	11,000	8	1.5	2.25
Bangladesh	23	4	1,810	12	−8	64
Iceland	14	11.5	28,270	4	7.5	56.25
Italy	9	15	31,810	3	12	144
UK	13	13	35,840	2	11	121
New Zealand	14	11.5	28,100	5	6.5	42.25
						Σ = 1,064.5

Source: figures from the *2012 World Population Data Sheet* (Population Reference Bureau, 2012)

Step 1 The formula for investigating Spearman's rank is:

$$r_s = 1 - \left(\frac{6\sum d^2}{n^3 - n}\right)$$

where d = the difference in rank of the values of each matched pair

n = the number of ranked pairs

$\sum$ = sum of

Step 2 The next stage is to work out the rank order of the values within each variable (from the highest value to the lowest). When there are two values the same, add both rank values together and divide by two. For example, Jamaica and Brazil both have a birth rate of 16 and they should be ranked 9 and 10 on the table. Add 9 + 10 and divide by 2 = 19/2, so both are ranked as 9.5.

Step 3 Calculate the difference (d) between each of the pairs of rank values (first value minus the second value).

Step 4 Square each of the resultant values (d^2). Note that all minus numbers should disappear at this stage.

Step 5 Add all the d^2 numbers to get the total $\left(\sum\right)$ for d^2.

Step 6 Go back to the formula and use the data to complete the calculation. Always remember to show your working so that you can score marks even if you get the wrong final answer.

$$r_s = 1 - \left(\frac{6\sum d^2}{n^3 - n}\right)$$

$\sum d^2$ = total/sum of the differences in the values of each matched pair squared

n = the number of different values measured for comparison (ranked pairs)

$$r_s = 1 - \left(\frac{6\sum d^2}{n^3 - n}\right)$$

$$r_s = 1 - \left(\frac{6 \times 1,064.5}{15^3 - 15}\right)$$

$$r_s = 1 - \left(\frac{6,387}{3,375 - 15}\right)$$

$$r_s = 1 - \left(\frac{6,387}{3,360}\right)$$

$$r_s = 1 - 1.9$$

$$r_s = -0.9$$

Step 7 Check that the final value lies between −1 and +1. If it does not then you have done something wrong in your calculation.

Step 8 Comment on the strength of the relationship — any result between 0 and −1 is seen as having a negative correlation. The closer the result is to −1, the stronger the correlation. Any result between 0 and +1 is seen as having a positive correlation.

The closer the result is to +1, the stronger the correlation. In this case, the relationship has a very strong negative correlation.

Step 9 Comment on the statistical significance. In the exam you will be given both a graph and a table to help you to work out the significance of your result (Figure 11).

When investigating correlation there is always the possibility of bias in collating the data or of a result occurring by *chance*. If there is a possibility that chance occurred within more than 5% of the data, then this is considered unacceptable and we should not accept the results. This is sometimes called the 5% rejection level.

Examiner tip

Spearman's rank is relatively straightforward when you know how. At least one of the statistical methods is likely to be examined at length in Unit 2. Know how to use them and, more importantly, know how to talk about the significance and how to generate a geographical reason to explain a correlation.

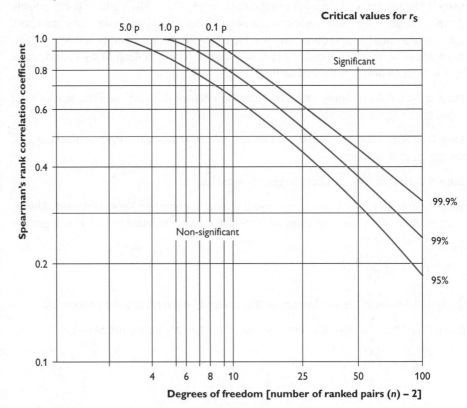

Figure 11 Spearman's rank correlation significance graph and table

Degrees of freedom	Significance level	
	0.05 (5%)	0.01 (1%)
8	0.72	0.84
9	0.68	0.80
10	0.64	0.77
11	0.60	0.74
12	0.57	0.71
13	0.54	0.69
14	0.52	0.67
15	0.50	0.65
20	0.47	0.59

Using the graph, plot the degrees of freedom on the horizontal axis by counting the number of pairs (n) minus 2. Then plot the Spearman's rank result on the vertical axis.

- If the value lands above the 99.9% significance level line, this means that the relationship is strongly significant and there is a probability of less than 1 in 1000 that the relationship occurred by chance.
- If the value lands between the 99% and 99.9% significance lines, this means that there is a probability of less than 1 in 100 that the relationship occurred by chance. We can still accept the relationship and result.
- If the value lands between the 95% and 99% significance lines, this means that there is a probability of up to 5 in 100 that the relationship occurred by chance.
- If the value lands on or below the 95% significance line, this means that there is a probability of at least 5 in 100 that the relationship occurred by chance and we find the result to be non-significant. We should reject any hypothesis, as the element of chance or bias interfering with the results is too much.

Step 10 Provide a geographical explanation — a final question part might ask you to 'give geographical reasons that could be suggested to explain this statistical result'. To answer this you need to go back to the original hypothesis and the original variables that are being tested and use the Spearman's rank result, strength and level of significance to show how strong the relationship is.

In this case, the result is a strong negative correlation (and strongly significant at the 99.9% level), which means that as one variable increases the other decreases. In this case, as the amount of GNI per capita increases the birth rate is expected to decrease.

Your explanation might then go on to explore reasons why people who live in a country where GNI is very high might then have enough money for contraception, or might be focused on careers instead of family size, with the result that birth rates are much lower than in countries with a lower GNI.

Nearest neighbour analysis

Many investigations involve the arrangement of spatial data, sometimes called 'measuring dispersions'. On maps, settlements often appear as dots. Nearest neighbour analysis (R_n) is a statistical measure that allows us to look at a distribution (usually a dot map) and to determine a pattern.

Nearest neighbour analysis produces a figure to show the extent to which the pattern on the map (Figure 12) is found to be clustered (nucleated), random or regular (uniform).

The basis of the statistic is to measure the distance between each of the dot points on a particular map.

> **Knowledge check 7**
>
> (a) Describe the statistical significance of the results.
>
> (b) Give geographical reasons that could be suggested to explain this statistical result.

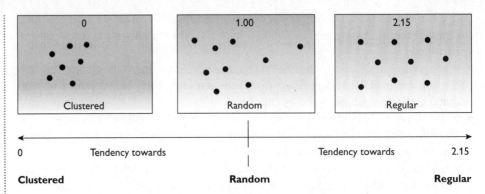

Figure 12 Patterns for nearest neighbour analysis

Step 1 The formula for investigating nearest neighbour is:

$$R_n = 2d\left(\sqrt{\frac{n}{A}}\right)$$

where *d* is the mean distance between the nearest neighbours, *n* is the number of points and *A* is the area under study.

Step 2 The first part of the formula is found out by using the map to work out which of the points is the nearest to each and to measure the distance between them. This should be measured in km. (See, for example, Figure 13 and Table 2.)

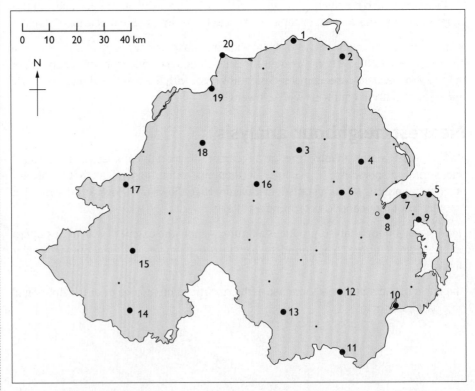

Figure 13 Map of Met Office automatic synoptic and climate stations, 2012

CCEA AS Geography

Table 2 Nearest neighbour measurements for Met Office stations in NI

Met Office station	Number	Nearest neighbour	Distance (km)
Giant's Causeway	1	2	20
Ballypatrick Forest	2	1	20
Portglenone	3	16	20
Killylane	4	6	12
Orlock Head	5	9	10
Aldergrove	6	4	12
Helens Bay	7	9	10
Stormont Castle	8	7	10
Ballywatticock	9	5	10
Murlough	10	12	20
Killowen	11	12	22
Katesbridge	12	10	20
Glenanne	13	12	25
Derrylin	14	15	22
St Angelo	15	14	22
Lough Fea	16	3	20
Castlederg	17	15	25
Banagher	18	19	20
Ballykelly	19	20	12
Magilligan	20	19	12
Map area = 165 km × 135 km = 22,275 km²			

Step 3 Calculate the mean distance (d) between the nearest neighbours by adding all the distances and then dividing by the number of points/places. In this case the total is 344 km, giving a mean of 344/20 = 17.2 km.

Step 4 Go back to the formula and use the data to complete the calculation. Always remember to show your working out so that you can get marks even if you get the final answer wrong.

$$R_n = 2d\left(\sqrt{\frac{n}{A}}\right)$$

$$R_n = 2 \times 17.2 \times \sqrt{\frac{20}{22,275}}$$

$$R_n = 34.4 \times \sqrt{0.0008}$$

$$R_n = 34.4 \times 0.028$$

$$R_n = 0.96$$

Step 5 Check that the final value lies lie between 0 and 2.15 — if it does not then you have done something wrong in your calculation.

Step 6 Comment on the pattern. On the exam paper you are given a graph to help you comment on the R_n value (Figure 14). A result towards 2.15 is regular while a result towards 0 is clustered. A result that is around 1 is random, but might have a tendency towards regularity or clustering.

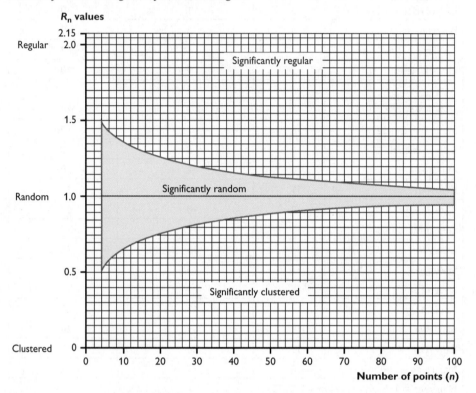

Figure 14 Nearest neighbour significance graph

Step 7 Comment on the statistical significance — the graph also helps you to work out the significance of your result. Plot the number of points along the horizontal axis and then read the R_n result up the vertical. This tells you if the result is significantly random, clustered or regular.

In this example, the distribution pattern of the weather stations is nearly perfectly, significantly random. There is a tiny tendency towards clustering but there is no real regularity at all. In many cases, a result that is random is considered to be at the 95% significance level, which means that there is a probability of 5 in 100 that the relationship occurred by chance, and so we reject the result and any hypothesis.

Step 8 Provide a geographical explanation — a final question part might ask you to 'give geographical reasons that could be suggested to explain this statistical result'. To answer this you need to explain why the location might be clustered, random or regular. In this example, the location of weather stations is random, though there is the slightest of tendencies towards clustering, which might be explained by the clustering of a few stations in and around Belfast.

Knowledge check 8

(a) What evidence might there be to show that this result is unreliable?

(b) What might be the impact on the R_n value if the area were changed to just look at the County Antrim area?

CCEA AS Geography

Limitations of using the nearest neighbour

Nearest neighbour analysis has a number of limitations that might create bias in the result:

- Mathematicians state that a minimum of 30 points is needed to ensure reliability and validity of conclusions.
- The size and scale of the area under investigation are extremely important. A larger area will lower the R_n value and will show more clustering. A smaller area will increase the R_n value and the pattern will be more regular.
- It is often used to measure settlements. Which settlements should be included? Do individual isolated buildings or hamlets count? For larger towns and cities there can be confusion about where the central area is, i.e. where the distance should be taken from.

Examiner tip
Nearest neighbour is often used for settlement studies, but be aware that it can have different uses. Make sure that you are aware of the different limitations of the technique.

Summary

- Photos and maps can be used for students to draw an annotated sketch.
- Dot maps can be used to illustrate distribution patterns in an area.
- Flow line maps can show the strength of a movement in a direction.
- Choropleth maps are visually appealing but can be difficult to draw.
- Isoline maps use lines to join areas with the same value.
- Bar graphs, line graphs, scattergraphs, proportional graphs and triangular graphs can all be used to present geographical data.

- Geographers must choose an appropriate sampling technique to collect their data (random/stratified/systematic/pragmatic).
- Mean, median, mode and range are simple statistical measures.
- Spearman's rank correlation measures the relationship between two sets of variables.
- Nearest neighbour analysis allows measurement of the pattern of dispersion in an area, ranging from clustered to regular spacing.

Topic 1 Population

Population data

National census taking

The Office of National Statistics (2010) notes the following:

> A census is a count of the population. We have to have one in the UK every ten years to find out more about who we are as a nation. We ask everyone to tell us a little bit about themselves to help census users decide how best to plan, fund and deliver the everyday services we all need — like housing, education, healthcare and transport.

Brief history of the census

Within the UK, a full census has been taken every 10 years since 1801 (apart from 1941). The amount of information collected has increased over the years, with many of the questions providing information on the population trends, resources and requirements for public services in an area. The census is seen as a **static** measure, as it takes a snapshot of the population of a country on a particular day.

The census in MEDCs

The most recent census in the UK (an example of a more economically developed country or MEDC) took place on Sunday 27 March 2011. In Northern Ireland the census was organised by the Northern Ireland Statistics and Research Agency (NISRA). This information can support government services in the following ways:

- Population — knowing how many people are living in an area helps the government allocate funding where it is needed.
- Education — the census helps to plan the location and need for particular education services.
- Health and disability — health services can be planned to make sure that healthcare is concentrated in the areas where it is most needed.
- Housing — housing needs can be better planned if the authorities know what the demand is now and what it will be in the future.
- Employment — information can be used to help plan jobs and training needs.
- Ethnic groups — census information can be used to help allocate resources and ensure that all groups are treated fairly.
- Transport — working out how and where people travel to work can help understand pressures on our transport system and improve public transport.

Knowledge check 9

Why is the census useful to national governments?

Questions on the census

The most important feature of the census is the range of questions that it asks.

Demographic questions: How many people live in your household? What age are they? How long have you lived in the UK? Were you born outside the UK?

Social questions: What passport do you hold? How would you describe your national identity? What is your ethnic group? What religion do you belong to? What is your main language?

Economic questions: How many cars or vans are owned, or available for use, by members of this household? What qualifications do you have? What is your main job? How many hours a week do you usually work? How do you usually travel to your main place of work?

Data collection and reliability issues

The census is a paper questionnaire that is delivered by post to each household across the UK. A team of 'enumerators' then collects the completed surveys in the weeks that follow before the answers are scanned into computers using optical readers. However, for the first time in 2011, householders were given a code that gave them the option of completing their census forms online. Over 99% of respondents returned their census form and a high level of quality assurance checking was carried out to ensure the validity of the completed forms.

The UK census is one of the most reliable in the world. However, issues can affect the reliability of the results.

- Government interference: some people, who might be claiming unemployment benefit whilst also working, might want to avoid the government knowing they are in employment and give false information.
- Confidentiality issues: some people do not want the government to have information on file about them and therefore refuse to fill in the form.
- Language, sight and special needs: some people find it difficult to read or to understand the questions on the form and might need help.

The census in LEDCs

In recent years, the United Nations has supported a large number of less economically developed countries (LEDCs) in improving their demographic data collection, but there can still be issues about the management and quality of data collection.

In a recent census carried out in Kenya on 24–25 August 2009, enumerators were trained to visit precise areas and count the number of people living in each. The census included questions on age, sex, tribe or nationality, religion, birth place, orphanhood, number of children, job, number of livestock, main source of water, main type of cooking fuel and access to different aspects of ICT (radio, television, mobile phone, landline and computer).

Data collection and reliability issues

Collecting the information across countless small villages and towns can be difficult and can compromise the accuracy of the data.

- Literacy levels: poor education rates in many LEDCs means that few people can read and write and would be unable to complete a written census form.
- Lifestyles: nomadic tribes and families can be difficult to find and might migrate across international boundaries.
- Size: large countries with countless scattered villages (e.g. India) can make the process of census organisation difficult.
- Fluid population: centres of mass migratory population like the shanty towns in São Paulo have a transient population, which is difficult to analyse.
- Cost: many LEDCs do not have the money to spend on such a task.
- Mapping inaccuracies: households might be left out if the mapping of areas is incomplete.
- Transport difficulties: it can be difficult for enumerators to gain access to some places, which can be made worse by seasonal rains or weather patterns.
- Cultural traits: in some areas of the Middle East, male enumerators are not permitted to interview women.
- Language barriers: tribal/ethnic languages can sometimes cause obstacles. In Cameroon, for example, there are more than 30 different language groups.
- Lack of reference: some people are not aware of what age they are, as they have no real point of reference from which to work this out.

Knowledge check 10

What are main issues of reliability for an MEDC census?

Examiner tip

Make sure that you understand the key issues behind census reliability in LEDCs and MEDCs. Issues with data collection practices in LEDCs are more obvious than in MEDCs, but questions on reliability issues can also be asked in relation to MEDCs.

Vital registration

Vital registration is a more *dynamic* aspect of population data, where the number of births and deaths within a country is monitored every day and is always changing. People are required to register births, still-births, deaths and marriages. This has been compulsory in the UK since 1847.

Registration of births

Usually a birth in NI needs to be registered within 42 days. Often information is collected within the maternity ward of the hospital and parents can then pick up a birth or adoption certificate from the General Register Office Northern Ireland (GRONI).

Registration of deaths

A death should be registered as soon as possible to allow funeral arrangements to go ahead and no later than 5 days from the date of occurrence except where the matter has to be dealt with by a coroner. The person registering the death must go to the Registrar with the medical certificate of the cause of death.

Vital registration measures are extremely reliable in MEDCs and mostly within LEDCs. Most LEDCs have some form of mechanism for registering the birth of a baby. However, there are some places where a written record of births is still not kept. There can be false claims for both births and deaths; this is more likely for deaths as associated family members might have more to gain from a certificate to say that a family member has deceased.

Vital registration reliability in Kenya

The reliability of vital registration data has improved within many LEDCs in recent years. In Kenya a campaign was launched in 2005 to improve the reliability of vital statistics. Up to that point only 34% of the rural population and 84% of the urban population were registered. A universal birth and death registration programme has now been implemented where parents must complete a notification form at the tribal chief's office when a child is born — but this process can still take up to 2 years.

> **Knowledge check 11**
>
> What might be a potential issue for the reliability of vital registration in an MEDC?

Summary

- A census is a survey of the population. In most countries this is carried out as a written or online questionnaire/form that heads of household must complete.

- In the majority of countries a census is carried out every 10 years.

- A census is a **static** measure of population as it takes a snapshot of the population make-up of a country.

- Each national census has questions that are specific to the particular needs of that country. In Northern Ireland, the census questions can be divided up into three distinct categories: demographic, social and economic.

- A census taken in an MEDC is likely to be at least 99.9% accurate. However, there can still be issues related to governmental interference, confidentiality and access to the census form.

- Vital registration is a dynamic aspect of population data where the numbers of births and deaths in a country are monitored.

- In most MEDCs vital registration is carried out at a local government level and reliability is close to 100%.

- In LEDCs there are issues of reliability as many countries have yet to develop universal registration practices and laws. In general, the more rural the community, the more difficult it can be to register a birth.

Population structure

Components of population change

You need to **understand components of population change**, and be able to **define crude birth rate**, **crude death rate**, **natural increase/decrease and migration balance**.

The population balance

The population of a country is dynamic and always changing:

births ± deaths = natural population change

The **crude birth rate** is the number of live births each year per 1,000 of the population in an area. Typically, birth rates are high in LEDCs (30–40 per 1,000) and low in MEDCs (8–15).

The **crude death rate** is the number of deaths each year per 1,000 of the population in an area. Typically, death rates are low in MEDCs (6–14) while in LEDCs some remain high (> 30) and others have fallen quickly (to around 8–10).

When there is a growth of population (a higher birth rate than death rate) we say that there is a **natural increase**. If there is a decline in the number of people (a lower birth rate than death rate) we say that there is a **natural decrease**.

In recent years one of the main factors affecting a population has been the amount of migration that takes place. Migration is the movement of people from one place to another. Migrants usually need to cross some form of administrative boundary for the migration to be counted in official figures.

The migration balance is the difference between in- and out-migration:

in-migration ± out-migration = migration balance

- **In-migration** (or **immigration**) is when people move into a country.
- **Out-migration** (or **emigration**) is when people leave a country.

> **Examiner tip**
> It is really important that you understand fully and are able to define the different elements of population balance. This is a common short question. Learn the definitions in detail.

Population pyramids

You should be able to **analyse population pyramids, including comparisons over time and space** (for *one* national case study).

How do population pyramids work?

A population pyramid is a graph that shows the age structure of a population, i.e. which age groups include more people than others. This helps us to understand what is happening within a country.

The horizontal axis shows either the total number or the percentage of males and females in the population. The vertical axis shows the different age cohorts. Usually this will be shown as individual age or 5-year cohorts (0–4, 5–9, 10–14 and so on).

You might have to decide whether a population pyramid is representative of an MEDC or an LEDC. An MEDC population structure (Figure 15) will usually have a much higher number of older people at the top of the pyramid (an aged dependent profile) whereas a LEDC population structure might have a higher number of younger people at the base of the pyramid (a youthful dependent profile).

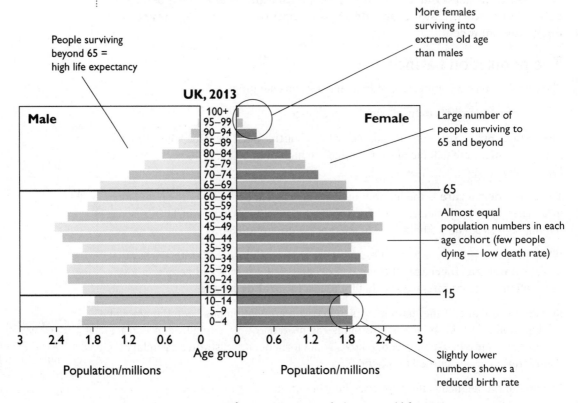

Figure 15 UK population pyramid for 2013

When looking at a population pyramid you should try to break the pyramid into three categories:

- youthful dependent population (aged 0–15)
- economically active population (non-dependents, aged 16–64)
- aged dependent population (aged 65 and over)

The number of people in each age cohort will help you discuss the birth rate, death rate and life expectancy in each country. The shape of the pyramid will often give clues as to the fertility and mortality changes that are taking place and the changing development of the country (Table 3).

Population pyramids can also reflect changes in the population caused by migration. A bulge in the structure might be found between ages 20 and 40, especially in more urban areas, as people move into an area. Equally, areas that experience out-migration (like rural areas) might seem to have a 'bite' out of the population structure.

Knowledge check 12

Explain the main components of a pyramid that are used to show how a population changes.

Table 3 Stages of development

Stage	Diagram	Comments
Stage 1 (LEDC)	Male (%) Female (%), 65, 15	High birth rate (wide base) High death rate (narrow top), with fewer people through each age cohort Very short life expectancy (around 30 years)
Stage 2 (LEDC)	Male (%) Female (%), 65, 15	High birth rate (wide base) Fall in death rate with slight increase in number of people surviving in each age cohort (top is widening) Still a relatively short life expectancy (around 40 years)
Stage 3 (LEDC/MEDC)	Male (%) Female (%), 65, 15	Falling birth rate (fewer children being born) Falling death rate (sides are becoming more straight) Life expectancy increases and more people are living beyond 65
Stage 4 (MEDC)	Male (%) Female (%), 65, 15	Low birth rate (narrow base) Low death rate (straight sides) Life expectancy continues to increase as people are living to 75 and beyond (top of pyramid is now wide)
Stage 5 (MEDC)	Male (%) Female (%), 65, 15	Very low birth rate (base is narrowing further) Low death rate Life expectancy continues to increase and there are more older people than young people

Population pyramids can change over space. A pyramid for one city might be different from that for another city or an urban area might be different from a rural area. Equally, pyramids can change over time as countries respond to demographic issues and challenges.

How population pyramids change over time: GB (England, Wales and Scotland)

Figure 16 and Table 4 show how the GB population has changed since the eighteenth century.

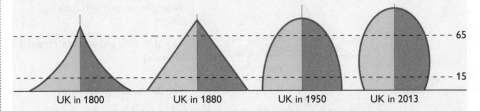

Figure 16 Estimated population structure for GB, 1800–2013

Table 4 Key facts relating to population structure change over time

Stage/period	Stage 1 1750–1800	Stage 2 1800–1880	Stage 3 1880–1950	Stage 4 1950 to present
Birth rate	High (37/1,000)	High (30/1,000)	Down (16/1,000)	Low (13/1,000)
Death rate	High (31/1,000)	Down (19/1,000)	Low (13/1,000)	Low (9/1,000)
Life expectancy	Low (45)	Improving (55)	Improving (65)	High (75)
Population (millions)	10.5	29	43	63

Explanation for the changes over time

Stage 1: 1750–1800

- Birth rates and death rates were high, but fluctuating, giving a small population growth.
- It was a youthful population structure.
- Birth rates were high because of:
  - a lack of family planning
  - a high infant mortality rate — parents had large families to ensure that some children reached adulthood
  - children were needed to work the land
- Death rates were high (especially among children) owing to:
  - the prevalence of disease/famine across the land
  - poor standards of living and hygiene
  - basic or non-existent medical care

Stage 2: 1800–1880

- Birth rates remained high as the food supply became more reliable.
- Death rates fell dramatically because:
  - mortality crises like plague or famine had been eliminated
  - of improvements in nutrition and standards of living
  - of improvements in medical care
  - infant mortality rates decreased due to improved healthcare

Stage 3: 1880–1950

- Death rates continued to fall and stabilise.
- The average woman was having 5.5 children in 1871 but this had fallen to 2.4 children in 1921. Birth rates fell quickly due to a number of factors:
  - Marriage was delayed or traditional methods of birth control were developed.
  - Lower infant mortality rate meant less need to continue to have children.
  - The Industrial Revolution was advancing rapidly through GB and increasing industrialisation meant that fewer workers were needed in urban factories.
  - There was increased desire for material possessions.
  - Improved roles and equality for women meant more women entered the workplace.

Stage 4: 1950 to present

- Birth and death rates became low but fluctuating.
- The rate of population growth slowed down quickly so that by 1960 the population was only growing by 5% each year and by the 1980s this had slowed to 1.9%.
- Demands for labour meant that GB had to search for migrant workers to keep the economy moving.

Examiner tip

This is a popular long question topic, so you need to make sure that you can describe how the population *structure* changes over *time*.

Knowledge check 13

Describe the way in which the GB population pyramid shape has changed, using birth rates, death rates and life expectancies to support your answer.

Case study

How population pyramids change over space: China

The total population in China was 1.3 billion in 2010. This was an increase of nearly 74 million (5.84%) since the 1990 census, an annual growth rate of 0.57% — showing that China had a much lower fertility rate than in the previous decades. Over 401 million households were counted, with an average of 3.1 people living in each house. Males accounted for 51% of the population with females accounting for 49%.

Migration within China continues to reinforce the imbalance between rural and urban areas. An increasing number of people are moving into the urban east of the country and away from the rural west. Roughly 94% of the population live in 40% of the living space with the remaining 6% living in 60% of the living space (Figure 17). This demographic divide within China means that different age profiles exist in the two areas.

Xinjiang is the most rural and most westerly province in China, sharing borders with India, Russia and Mongolia. It is a large, sparsely populated area of around 1.6 million km². Xinjiang takes up about a sixth of the Chinese living space. It currently has a population of 21 million people, giving a population density of 13 per km². It is known for its agricultural produce, including grapes, melons, pears, cotton, wheat, silk, walnuts and sheep.

Guangdong is an urban area in the southeast of the country, which includes the major cities of Hong Kong and Macau. It is a relatively small area in China, around 180,000 km², with a population of 104 million people, giving a population density of around 580 per km². It is currently the most productive industrial and financial region in China, with the highest GDP figures.

Figure 18 shows the population pyramids for Xinjiang and Guangdong.

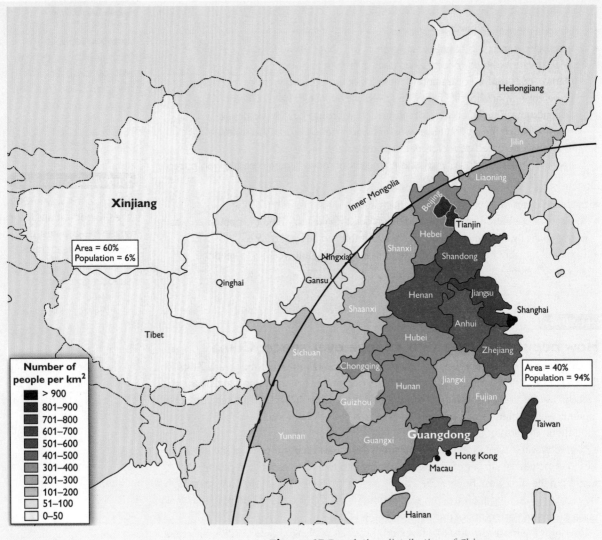

Figure 17 Population distribution of China

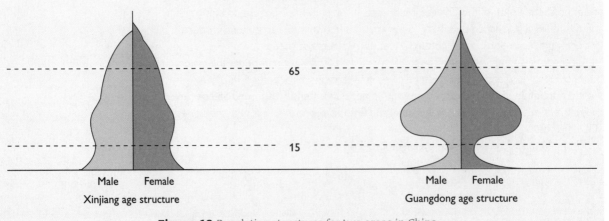

Figure 18 Population structures for two areas in China

Explanation for the changes over space

Xinjiang

- The birth rate is stable. There are fewer young people aged 0–14 than in Guangdong. The birth rate is 15 per 1,000.
- 15–39 is the most represented age range in the population. There is a slight imbalance as more men stay to work the land, while some women migrate to the urban areas. Work is hard in the remote rural regions.
- Life expectancies are quite low. Few of the population reach old age due to the harsh living conditions and subsistence farming. However, women survive longer than men.
- There is some migration as men move into the area to work on the land while some young women migrate to the more urban areas of China.

Guangdong

- Birth rates in the urban areas have started to rise a little in recent years for two reasons:
  - As standards of living have risen, with salaries growing in urban areas, people can afford to pay government fertility fines for second and third children.
  - The massive population is made up of mostly young people who are in their reproductive prime, and they want to start families.
- However, birth rates are still low compared with the rest of the country (around 8 per 1,000).
- The working population (especially those aged 15–24) form the biggest age categories. There is a huge influx of labour into the factories and offices located in the area. The age category of 20–24-year-olds makes up 8% of males and 10% of the whole population. Many of the young women are sought after to work in the 13,000 processing, assembling and manufacturing industries in the area.
- There is a huge amount of migration into the area on both a permanent and temporary basis. Every year an estimated 30 million workers (the 'floating population') migrate to work in the cities and towns for about 6 months of the year before returning to their rural lives.

> **Examiner tip**
>
> This is another popular long question topic. Make sure that you can describe how the population *structure* changes over *space*.

> **Knowledge check 14**
>
> What are the key facts that you need to remember about each of these regions?

Economic, social and political implications of dependency ratios in MEDCs and LEDCs

Dependency ratios

In any country there are people who contribute to the tax burden of the country and those who do not. Dependency ratios help to demonstrate the balance between those in work (the working population) and those who are not (the young and elderly who do not pay taxes).

$$\text{dependency ratio} = \frac{\text{youthful dependent population (0-14)} + \text{aged dependent population (65+)}}{\text{working population (15-64)}} \times 100$$

MEDCs usually have a dependency ratio of between 50 and 75. This means that 100 people need to work in order to pay for 50–75 dependents. In LEDCs, the dependency ratio is usually much larger than 100.

The usual pattern in MEDCs is an increasing number of aged dependents but a decreasing number of youthful dependents. In LEDCs the youthful dependents outnumber the aged dependents.

In 2011 the dependency ratio for the UK was 51.87; for Kenya it was 82.14.

Knowledge check 15

Describe how the dependency ratio works.

The economic, social and political implications of an aged dependency in MEDCs

The demographic trend in many MEDCs shows the number of youthful dependents to be decreasing and the number of aged dependents to be increasing. For example, in the UK the birth rate is 12/1,000 and the death rate is 8/1,000. An increase in the number of old people has the following implications.

Economic

- **Healthcare:** people who live longer require expensive healthcare and treatments. The government needs to spend increasing amounts of money just to keep people alive.
- **Residential care:** as aged people mature they start to lose their independence and might need additional residential accommodation.
- **Increased pension burden:** in the 1970s, when life expectancies were 10 years lower, people would draw their pension from 60 but die at 67. Today, a woman might retire at 60 and live to 85 — drawing a pension for 25 years. This has created great pressure on the government.
- **Disposable incomes:** many aged people find that they are much better off than they were in their working years.

Social

- **Impact on the family:** families are extending. Grandparents are alive for longer, which can be a good thing because they can help with childcare responsibilities, releasing both parents to return to work.
- **Degenerative illnesses:** as people live longer, so many suffer from degenerative illnesses such as Alzheimer's or Parkinson's, which can put pressure on the family unit.
- **Loneliness:** there is often a difference of about 7 years between the life expectancies of women and men. Many women are widowed early and have to spend their last years alone.
- **Ageism:** many firms have employment rules that do not allow people to work beyond the age of 60. This makes it hard for them to get a full pension and it is difficult to get a new job at this age.

Political

- **Pension decisions:** the government needs to make sure that it has the right amount of money coming into and going out of the economy through pensions.
- **Paying for care:** the government makes decisions about how little money a person should have before their care is paid for by the government.
- **Euthanasia policies:** in some countries, people are pressing governments to make it easier for people to decide when and how they are allowed to die. Right to die and euthanasia policies are debated by governments regularly.

The economic, social and political implications of a youthful dependency in LEDCs

The demographic trend in many LEDCs is for the number of youthful dependents to be increasing at a fast rate. For example, in Kenya (Figure 19) the birth rate is 35/1,000 and the death rate is 8/1,000.

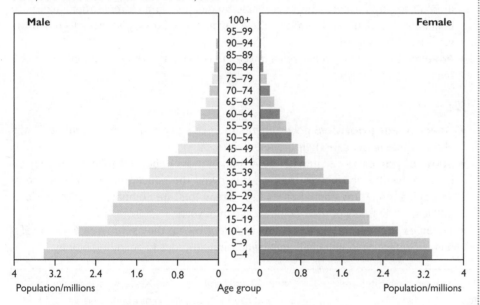

Figure 19 Kenyan population pyramid for 2013

An increase in the number of young people has the following implications.

Economic

- **Taxes and government income:** many of the governments in LEDCs find it difficult to raise money through taxes, which means that they have little money to be able to use to invest in services to support the population.
- **Education:** education authorities in many LEDCs have very little money to invest in the creation of a quality education system. Children must pay for their education and teachers are paid a minimum wage. With the increase in number of children accessing the services, these education systems have been pushed to breaking point.

Explain the expected economic, social and political implications for an ageing population structure.

- **Healthcare:** the government cannot afford to sink large amounts of money into medical care. People have no money to pay for healthcare and as a result they die from preventable illnesses and death rates remain consistent.
- **Informal employment:** it is difficult for people living in the slums and shanty towns in LEDCs to get formal jobs. More people moving into the cities put increasing pressure on these jobs and competition for places means that factory owners can get away with paying workers less money or forcing them to work in unsafe conditions.

Social

- **Education:** the population in Kenya is growing fast (around 2.7% per year) and this puts pressure on resources for young people. The government has relatively little money to spend on schools so few children have the opportunity for an education and are unable to read or write.
- **Medical issues:** people in LEDCs often die from preventable illness. It can be difficult to pay for even the most basic medicines, so children suffer. Access to hospitals is difficult and families often have to travel long distances to reach the most basic healthcare.
- **Poverty:** many children are born into a life of poverty, which is difficult to break out of.

Political

- **Government priorities:** governments in LEDCs need to make careful decisions about where to allocate their limited resources.
- **Show of power:** governments often think it more important to spend money on icons of wealth and power than on improving the education and healthcare of its population. Many capital cities have areas that are comparable with those in MEDCs, while many of their children live in poverty.
- **Voter expectations:** as the young people grow up they soon create a powerful base of voters, who will challenge traditional attitudes to spending money.

> **Knowledge check 17**
>
> Explain the expected economic, social and political implications for a youthful population structure.

Summary

- Population increase occurs when birth rates are higher than death rates.
- Population decrease occurs when death rates are higher than birth rates.
- Population pyramids show how the population structure of a country is changing either over time or space.
- The shape of the pyramid is dictated by the BR, DR and LE for that population.
- The population structure for GB has gone through many changes. The DR and BR were both high, then DR fell, followed by BR, and now both BR and DR remain low but fluctuating.
- In China, rural areas like Xinjiang have a different population structure from urban areas like Guangdong. Migration has an important part to play in each population.
- The dependency ratio for an area or country demonstrates the balance between those in work and those who are not.
- Many MEDCs have an aged dependency structure, which has economic, social and political implications.
- Many LEDCs have a youthful dependency structure, which has economic, social and political implications.

Population and resources

You need to be able to explain the relationship between population distribution and resources (for *one* national case study of population distribution — France).

Population distribution is the spread of people across an area. Where people are concentrated into an area we say that the area is densely populated. When they are spread out we say the area is sparsely populated.

Resources are defined as being any aspects of the environment that are used to support human needs.

Table 5 Some of the major factors affecting population distribution

Factors	Densely populated areas	Sparsely populated areas
Relief (physical)	Flat, low-lying river valleys are favoured	Mountains with low temperatures and high amounts of precipitation make it difficult for people to survive
Climate (physical)	Reliable but not extreme rainfall; good temperatures — neither too hot nor too cold — with a long growing season	Either very low or very high amounts of rainfall, which can lead to flooding; very hot or very cold areas
Vegetation (physical)	Areas with grassland tend to encourage more people to live there	Places with no vegetation or complex vegetation like forests have fewer people living there
Soils (physical)	Deep, fertile soils — especially those in river valleys	Places with thin or frozen soils; areas liable to soil erosion
Water (physical)	A good water supply is needed for drinking, sewerage, washing, farming and industry	Fewer people live in areas where there is no regular water supply
Pests and diseases (physical)	Fewer pests and diseases as a result of climatic factors	Many pests and diseases, restricting the lives of people who there
Resources (physical)	Mineral deposits and energy supplies allow large-scale manufacturing and support dense populations	Fewer resources means a sparse population
Communications/ transport (human)	Areas where it is easy to build canals, railway lines, roads and airports will attract settlement	Areas where it is difficult to build transport systems become inaccessible
Economic/jobs (human)	Areas with much farming activity or industry will attract people	Areas with less farming or industry offer few job opportunities for people
Political (human)	Government decisions might influence the dispersion of people around a country, e.g. creating new towns and cities in rural areas	Peripheral areas often receive less funding and investment and therefore stagnate, driving people away

Examiner tip

It is useful to consider the different physical and human factors, including resources, that influence a population *before* looking at a case study.

France

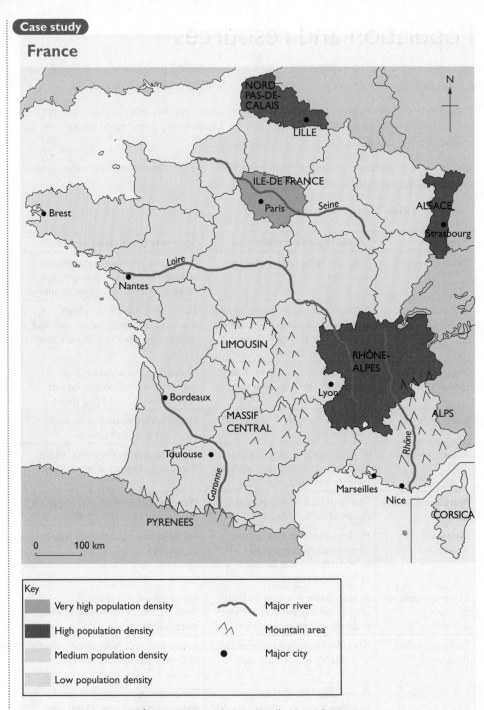

Figure 20 Population distribution of France

France has an interesting population distribution pattern (Figure 20). Some areas, such as the Massif Central, Southern Alps, Pyrenees, Corsica, Limousin, Auvergne and Aquitaine, are sparsely populated. Other areas, like the Ile-de-France (Paris), Rhône-Alpes, Provence, Toulouse, Brittany and Nord-Pas-de-Calais, are more densely populated.

Since the middle of the nineteenth century there has been a depopulation movement from the rural areas towards the urban areas (and Paris, in particular). The population of France in 2013 was 64 million and only about 10 million, or 15%, were continuing to live in the rural areas. The population density is around 117 people per km^2, which is relatively high for Europe.

Sparsely populated areas (e.g. Limousin)

The Limousin area of France contains around 742,700 people occupying some 17,000 km^2. It is the second least populated region in France and has a population density of 42 people per km^2.

The majority of the area is located in the Massif Central. This mountainous area, with a highest peak of 1,886 m, has many physical factors to deter population. The climate is often warm and wet in the summer and cold and wet, with snow and ice, in the winter. Soils are often thin and cannot support even the most basic vegetation.

There are relatively few energy resources in the area, though in recent years wind farms have been built on the higher land.

Densely populated areas (e.g. Ile-de-France, Paris)

The Ile-de-France/Paris area of France contains around 12 million people on 2,844 km^2. It is the one of the most heavily and densely populated areas in Europe — central Paris has a population density of 22,000 people per km^2.

The majority of the area is located in the Seine river basin. The Seine is one of the biggest rivers in France and Paris is located on a relatively flat part of the basin. The main population of Paris lives within a space about 87 km^2, surrounded by the forests of Boulogne and Vincennes.

The climate is mild and moderately wet, but summer days are usually warm. Winters remain mild, with frost and snow being uncommon. Rain is fairly regular and consistent throughout the year (with around 600 mm per annum). The area has nutrient-rich soils that are good for farmland (especially arable crops).

Much of the communications system within France starts in Paris. The canal, rail and road networks all have Paris at the central hub, which means that most journeys across France require a trip through or around the capital. The fast trains and autoroutes make travel in Paris fast and efficient.

Much of the French economy is based on Parisian industry. The GDP and income per capita in Paris are among the highest in the world. In 2009, for example, the GDP per capita was $46,800 for over 10 million people.

Since before the French Revolution in the eighteenth century, Paris has been the central focus for the French population. There was a steady migration stream from all rural areas towards Paris, further accentuated by the Industrial Revolution.

Urbanisation has also been the main feature of modern France. Paris continues to expand, with an estimated 22% of residents accounted for by international immigration. Paris has a more youthful age structure than other regions in France, with many retired Parisians moving to the south of the country.

The government has tried to stem the tide of people moving into Paris. Through the 1990s and early 2000s it decentralised many different governmental functions to more peripheral cities around the country. However, although moderately successful, this did not deter those who wanted to migrate to the economic hub.

Examiner tip

Make sure that you understand and can clearly identify the different physical and human factors that led to some areas being sparsely populated while others became densely populated. Facts and figures are vital for high marks in questions on this topic.

Summary

- Population distribution is the spread of people across an area. Areas can be densely or sparsely populated.
- Factors affecting population distribution are relief, climate, vegetation, soils, water, pests and diseases, resources, communications/transport, economics/ jobs and politics.
- The most densely populated area in France is central Paris, with a density of 22,000 people per km².
- One of the most sparsely populated areas in France is Limousin, with a density of 42 people per km².

Topic 2 Settlement and urbanisation

Challenges for rural environments

Issues in the rural–urban fringe

The rural–urban fringe is the hinterland between a typically urban landscape and a typically rural landscape.

Figure 21 shows that rural areas are very different from urban areas. As you move away from the CBD in a city, the amount of rurality increases.

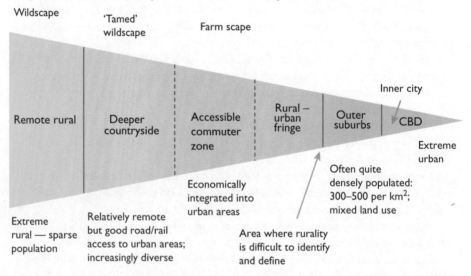

Figure 21 The rural–urban continuum

The extent to which an area is defined as rural or urban depends on a number of key factors:

- **Settlement size:** many countries have set criteria of how many people need to live in an area for a particular designation. In the UK, a settlement with a population of less than 1,000 is usually defined as being rural.
- **Population density:** an area with more than 100 people per km² might be described as being urban, while one with less than 100 would be defined as rural.

CCEA AS Geography

- **Settlement function/land use/employment:** sometimes the function of an area will help to show whether it is more indicative of a rural or urban population.
- **Perception and service provision:** it is generally accepted that rural areas are more likely to be characterised as having a more dispersed population, more agricultural or other extensive land use, and to be further away from major urban centres.

Table 6 shows an 'index of rurality', developed by Professor Paul Cloke, which identifies some indicators of rural life using available census data.

Table 6 Cloke's index of rurality

Indicator of rurality	Application to rural area
Population per hectare	Low
Percentage change in population	Decrease
Percentage of total population over 65	High
Percentage of total population male, 15–45 years old	Low
Percentage of total population female, 15–45 years old	Low
Occupancy rate: percentage of population at 1.5 people per room	Low
Households per dwelling	Low
Percentage in socio-economic groups 13/14/15, associated with farm work	High
Percentage resident for less than 5 years	Low
Distance from nearest urban centre	High

Greenfield developments

In the UK the growth of the 'suburbs' started in the 1920s. This is sometimes called 'urban sprawl'. In many towns and cities, new developments and housing estates are built at the edge of cities to reduce the pressure on the inner city. Land use planning around the edge of urban areas is critical to ensure that urban sprawl is contained and that land is used in the most effective way.

Greenfield development refers to an area of land surrounding a city or town that has not been developed or built up. Green belt policies are often put in place to actively prevent urban sprawl into these areas. Planners use the concept of a green belt as an invisible line used to prevent urban development into an area, seeing it as the 'lungs' of an urban landscape.

Belfast in 1971 had a population of around 600,000 and over the next 20 years there was a slow but steady decline of the inner city population as people moved to the suburb areas to the north, east and south of the city. In 1964 there was an attempt to stop the continued sprawl of Belfast with the Matthew Regional Plan. Matthew noted the need for a 'greenscape', capable of sustaining agriculture, forestry and outdoor recreation, that was convenient to the urban population.

Green belts have been quite successful in slowing urban sprawl. However, there have been a number of side effects:
- Green belts sometimes force development to take place further into the countryside.
- Inner-city areas can become more densely packed.

Examiner tip

It can be difficult to describe the differences between rural and urban areas, so make sure that you have a clear understanding of how this works.

Knowledge check 18

Describe how greenbelts might stop urban sprawl.

- Competition for land increases within the city, forcing land prices up (and in areas surrounding the greenbelt).
- Longer commuting distances into city centres increase congestion and pollution, resulting in a need for improvement to the transport network.

Suburbanisation

The process of suburbanisation is often the first step in the decentralisation of the inner city. In 1964, Matthew commented that Belfast needed some of its slum areas cleared and proposed measures to move people to the suburbs and beyond. Much of the subsequent suburbanisation began as extensions and improvements to the transport network and infrastructure were made (Figure 22).

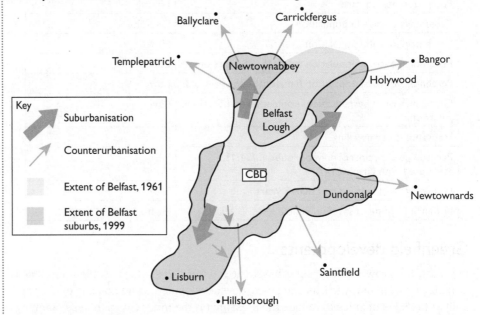

Figure 22 Suburbanisation and counterurbanisation around Belfast

As a city grows outwards and encroaches on the land surrounding it, it changes the characteristics of the urban landscape. The developing suburban areas experience pressure as more people create more demand for services, housing, shops, parking and social facilities. Sometimes these areas expand to merely become 'dormitory towns' or 'commuter villages' from where people travel to work.

Counterurbanisation

If suburbanisation is the movement of people from the inner city to the suburbs, then counterurbanisation is the movement of people from the inner city and suburbs to places beyond the city limits/metropolitan area. This usually involves movement to surrounding rural towns and villages that are within commuting distance of the city.

Since the early 1980s, people have continued to move out from the fringes of Belfast. These movements occurred as incomes allowed people to consider living in bigger, more expensive houses, with land and space around them.

Towns like Ballyclare, Templepatrick, Comber, Carrickfergus, Bangor, Newtownards, Greenisland, Moira and Hillsborough have all grown as a result of this counterurban movement.

Transport infrastructure

It would be difficult for any suburbanisation or counterurbanisation to take place without good transport linkages to the places people are moving to. Belfast already had a well-developed road and rail network, so people could choose to use their own independent transport or the public transport system.

What impacts have these developments had?

The above movements have led to a number of key issues within the rural–urban fringe, including:

- traffic congestion and pollution
- competition for land raising house prices
- the need to build and develop similar services to those in the city (e.g. shops, restaurants, entertainment)
- continued sprawl into the countryside — removing green space and reducing habitats for wildlife

Issues in remote rural environments

Places that are found far away from main centres of population can sometimes experience unique issues. A remote rural environment usually refers to a place that is far enough away from urban areas so as not to be affected by suburbanisation or counterurbanisation.

Population change: out-migration and an ageing population

The general trend in these areas is for any remaining young people to move out of the area towards local urban centres of population. They notice the decline in the rural area and want to be able to escape from their lifestyle, moving to the nearest city at the earliest opportunity.

Increasingly, the rural population gets older and fewer young people are left to support the population. In the Outer Hebrides, which forms part of the Scottish Highlands and Islands, population change has been a constant factor due to land clearances and reform, as well as changes to agriculture, the economy, cultures and the aspirations and expectations of people (especially the young).

Between 1901 and 2001 there was a 43% decline in the population of the Outer Hebrides. On the island of Barra, the population decreased from 2,545 people in 1901 to 1,172 in 2001 (–54%) and that of the island of Uig from 4,497 in 1901 to 1,527 (–66%).

Service provision: access to jobs/transport/communications/ services

Remote rural areas by their very nature have experienced a decline in the jobs available. Changes to the technological aspects of agriculture over the last 30 years mean that there are fewer employed on farms that at any other time. Transport infrastructure is

Examiner tip

Students often get the urban movements mixed up, so make sure that you read the question carefully before explaining the process.

Knowledge check 20

What impact does urban sprawl have on the rural–urban fringe?

Knowledge check 21

What potential effect might urban sprawl have on the city centre/inner city?

Knowledge check 22

Explain why out-migration can leave an aged population in an area.

Examiner tip
You need to make sure that you can make some 'general reference to places for illustration' when talking about issues in remote rural environments. Make sure that you have learned specific issues and can discuss them fully.

limited and does not allow easy access for people or goods. Communications can be basic and services are much poorer than in urban areas.

As people move out of the area, services are further weakened and disappear as the threshold to make them financially viable disappears. Local services like village shops, public houses, schools, post offices, churches, GP practices, police stations and local bus services have been in stark decline in recent years. Even simple things like broadband connections can be problematic and people begin to feel increasingly isolated and 'left behind' (Figure 23).

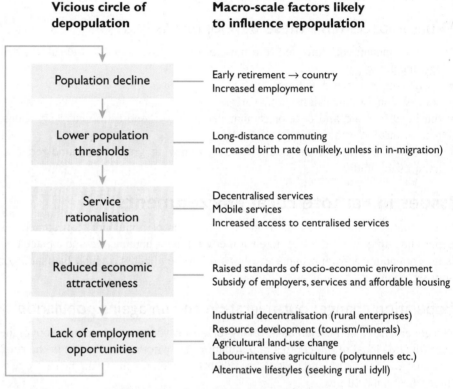

Vicious circle of depopulation	Macro-scale factors likely to influence repopulation
Population decline	Early retirement → country Increased employment
Lower population thresholds	Long-distance commuting Increased birth rate (unlikely, unless in in-migration)
Service rationalisation	Decentralised services Mobile services Increased access to centralised services
Reduced economic attractiveness	Raised standards of socio-economic environment Subsidy of employers, services and affordable housing
Lack of employment opportunities	Industrial decentralisation (rural enterprises) Resource development (tourism/minerals) Agricultural land-use change Labour-intensive agriculture (polytunnels etc.) Alternative lifestyles (seeking rural idyll)

Figure 23 The cycle of rural depopulation

Summary

- The rurality of an area depends on settlement size, population density, settlement function/land use, employment, perception and service provision.
- It is getting increasingly difficult to tell where urban areas stop and rural areas start.
- A green belt can be used to protect the land around a city and will stop the city encroaching further.
- Suburbanisation is the movement out of the inner city to the city suburbs. It is mostly caused by social and economic change within the city.
- Counterurbanisation is when people move from the inner city and the suburbs to places out of the city limits/metropolitan area. This is facilitated by having a good transport network.
- Remote rural areas are experiencing massive out-migration as young people leave to find work in the cities. This is having a big impact on the provision of services and access to jobs, transport and communications.

Planning issues in rural environments

Countryside management for conservation, recreation and tourism

You need to be able to explain how the countryside is managed for conservation, recreation and tourism, for example through Areas of Outstanding Natural Beauty (AONBs), National Parks and Sites of Special Scientific Interest (SSSIs).

An **AONB** is described as being an area of the countryside that is considered to have landscape of significant value. It is a term used in England, Wales and Northern Ireland and its main purpose is to conserve and enhance the natural beauty of the area. AONBs usually have a series of planning controls that help manage the area, whilst allowing people to enjoy it. There are 33 AONBs in England and Wales and nine in Northern Ireland (the most famous being the Antrim Coast and Glens, the Causeway Coast, the Mourne Mountains and Strangford Lough).

National Parks are managed carefully for conservation purposes. The term 'National Park' is an international designation covering around 6,555 parks worldwide. In the UK, National Parks are seen to have two central purposes: to conserve and enhance the natural and cultural heritage of the area and to promote understanding and enjoyment of the special qualities of the park by the public. There are 15 National Parks in the UK (10 in England, three in Wales and two in Scotland).

An **SSSI** is a conservation designation used in the UK to help protect some form of geologically or biologically significant area. The protection offered by designation as a 'Site of Special Scientific Interest' is usually for a small, localised area. In Northern Ireland this is controlled by the Northern Ireland Environment Agency (NIEA), which intends to have 440 SSSI areas declared by 2016.

Examiner tip

Make sure that you have a good knowledge of the different countryside management designations.

Managing the countryside for conservation

The main aim of these protection measures is to ensure that the countryside is conserved for the future. This is done by preserving or protecting an area and making sure that any natural resources are used wisely. This means that planners within a rural area need to consider carefully the ways in which the area can be used and need to make sure that any development is sustainable. Measures need to be taken in planning land use and recreation/tourism activities to make sure that the fabric of the landscape is not changed beyond all recognition.

Managing the countryside for recreation

As the number of visitors and the range of activities increase, the people who manage the area need to establish its **carrying capacity**. This is the level of use that a resource or an area can support before it starts to suffer significant deterioration.

Recreation is when people use their leisure time for enjoyment or pleasure and do not stay overnight, for example a day trip to go walking. These activities often attract

visitors to rural areas and provide much needed jobs and income for local people. For this reason, a trade-off between using the landscape as a resource and protecting the landscape needs to be found.

Managing the countryside for tourism

There is much overlap between recreation and tourism. Tourists are drawn to an area because of its recreational resources. However, tourists may travel to an area and stay overnight for recreational, leisure or business purposes. This means that tourists require services and facilities that allow them to develop a 'home from home'.

Option 1 Case study

The management of one area for conservation, recreation and tourism in one protected area (Peak District National Park)

The purposes for National Parks (in England and Wales) are as follows:
- Conserve and enhance the natural beauty, wildlife and cultural heritage.
- Promote opportunities for the understanding and enjoyment of the special qualities of National Parks by the public.

This case study relates to the Peak District National Park (Table 7).

Table 7 Key facts about the Peak District National Park

National Park designation	1951
Population	38,000
Area	1,437 km^2
Main settlements	Bakewell and Tideswell
Ancient monuments	457
Conservation areas	109
Visitors per year	8.4 million
Visitor days per year	10.4 million
Visitor spend per year	£356 million

Management for conservation

The challenges

Soil and hillside erosion: hill walking, mountain biking, quadbiking and 4×4 off-road driving have all become popular in the park. Heavy use of paths can cause extensive damage to soil and hillsides. Overgrazing by sheep on hill areas can also reduce vegetation, leading to soil erosion.

Damage to wildlife: erosion as well as clearance of hedgerows to enable the extension of farmland or recreational facilities destroys wildlife habitats, including bird nesting sites. Litter can be a danger to animals.

Damage to farmland: walkers might leave gates open and livestock can be scared or attacked by dogs.

The strategy

Protecting the landscape: the Peak District National Park Management Plan protects the landscape by working with farmers and other land managers to encourage land management systems within the park. The character of the park is to be enhanced. Heritage, historic buildings and archaeological remains are given statutory protection. A strict protection plan will be put in place in the Natural Zone.

Protecting the biodiversity and ecosystems: the diversity of wildlife has been under threat due to the rapid rate of change, so the countryside is being managed more carefully. Healthy soils and watercourses are vital, so these are being maintained in high-risk areas. Local partnerships with protection organisations like the National Trust and the RSPB are being enhanced.

Management for recreation

The challenges

Congestion of villages and beauty spots: some of the most popular areas, known as 'honeypots', attract large numbers of visitors. This can lead to overcrowded car parks, blocked roads and overstretched local resources.

Impact of recreational activities: some activities cause damage and create conflict. For example, high-speed boats can cause excessive amounts of noise pollution and might conflict with fishing or swimming.

The strategy

Sustainable approach to transport: cycle routes have been developed but occasional visitors are not using these to get into the park area. More sustainable local transport links into local towns and urban centres are being encouraged.

Management of recreational activities: many of the activities are closely managed and controlled. Watersports are limited to only a few of the reservoirs and other waters, while fishing is controlled through the involvement of fishing clubs. Motor sport activities (such as off-road/4×4 driving) are limited to particular areas, while traffic-calming devices are being fitted to 'green lanes'.

Management for tourism

The challenges

Impact on local services: services opened up to serve tourists, such as gift shops and cafes, can displace those serving the local population, including bakers and butchers. House prices rise as demand for second homes and holiday cottages increases.

Pressure on services: tourists require new hotels, holiday cottages and caravan parks. These take up valuable living space in the local area. Improved sewerage, electricity supplies, water supplies, phone lines and internet connectivity are all required to service the tourists.

The strategy

Sustainable tourism: the Management Plan aims to welcome people for 'escape, adventure, enjoyment and sustainability'. This involves management of the natural environment, heritage assets, local culture and local infrastructure so that the specific needs of visitors can be met.

Sustainable communities: the aim is for revenue from tourism to allow sustainable development within the local communities so that small settlements can adapt to new challenges, but still regain their historic and cultural identity.

Knowledge check 23

Describe the different ways in which the Peak District National Park can be managed for conservation.

Knowledge check 24

Describe the different ways in which the Peak District National Park can be managed for recreation.

Knowledge check 25

Describe the different ways in which the Peak District National Park can be managed for tourism.

Economic regeneration in remote rural areas

You need to be able to explain how economic regeneration is delivered to remote rural areas by regional development agencies.

Regional development agencies (RDAs) were set up in the UK in 1998 to stimulate economic growth in different areas across the country. Nine separate bodies were set up across England, with the Scottish, Welsh and Northern Irish governments setting up their own regional departments. The RDAs were abolished in March 2012 with local councils tasked to take on a similar role, but with a lot less finance than before.

Option 2 Case study

A regional development agency in a remote rural area (the Highlands and Islands Enterprise)

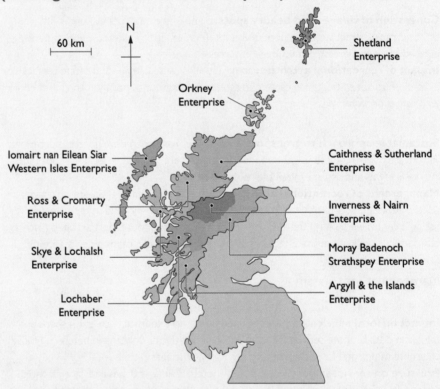

Figure 24 The HIE area in Scotland

The Highlands and Islands Enterprise (HIE) was set up in 1991 (Figure 24). The vision for the area is to create a highly successful and competitive region where increasing numbers of people choose to live, work, study and invest. The main aim of the HIE is to build and support the communities throughout this area and enhance their economic opportunities.

How does the HIE promote regeneration?

Regeneration through the work of the HIE aims to make the different remote rural communities more sustainable and self-sufficient. Some of the HIE's strategies for regeneration are shown in Table 8.

Table 8 The HIE's strategies for regeneration

Business support	The HIE actively helps small, local businesses to export their goods to the international market (it created 130 new jobs in 2012–2013). It has a forward-thinking innovation and R&D unit, which helps product development.
Community support	The HIE has set up a special range of resources aimed at supporting small communities, helping to manage and maintain community assets, community funds and encouraging community approaches to renewable energy and crofting (agriculture).
Growth sectors	The HIE has been investing in traditional local industries (e.g. beef farming, fishing, and the manufacture of shortbread, oatcakes and malt whisky), but it has also branched out into new areas (£4.9 million was spent supporting 77 projects in 2012–2013). One of the UK's newest universities — the University of the Highlands and Islands — was set up across 50 outreach centres and caters for 7,500 students. Tourism continues to be an important source of revenue, for example from the Isle of Skye, Loch Ness and the ever expanding whisky trails through the islands. The aim is to grow the tourist sector by £1.5 billion over the coming years.
Investing in a renewable future	There have been big investments in renewable energy throughout the region, including the building of the European Marine Energy Centre in Orkney in order to research wave and tidal energy. Investment in the Nigg fabrication site in Invergordon will engage over 2,000 people in a modern energy park where research into hydro power and offshore wind will enhance current renewables projects.

Examiner tip
Make sure that you know and can explain the aims of the HIE in regenerating the economy in this area.

Examiner tip
Make sure that you can explain the different strategies used to promote economic regeneration. How successful have they been?

Summary

- Rural environments need protection and various planning mechanisms are used to do this, including AONBs, National Parks and SSSIs.
- The countryside is managed for conservation, recreation and tourism.
- The Peak District National Park is an example of an area that has developed an effective management plan to deal with conservation, recreation and tourism.
- The government often needs to intervene in remote rural areas and has done this through the organisation of regional development agencies.
- The Highlands and Islands Enterprise in Scotland is a good example of an RDA that has been involved in regenerating the economy of a remote rural area.

Challenges for the urban environment

Issues of the inner city in MEDCs

You need to be able to demonstrate knowledge and understanding of **issues of the inner city in MEDCs, including social and economic deprivation, re-urbanisation and gentrification**, with a case study to illustrate these urban issues (e.g. Belfast).

Case study

Belfast

Belfast grew rapidly in the 1800s with the introduction of textile and linen mills along the rivers of north and west Belfast (e.g. on the Shankill, Crumlin and York Roads). Engineering works (e.g. Mackies and Sirocco) grew up along the Springfield and Newtownards roads and shipbuilding and its associated industries (like rope-making and furniture making) developed near the mouth of the River Lagan (Figure 25).

Inner-city Belfast continued to grow rapidly. Much of the housing was built close to the factories in long, straight terraced formations.

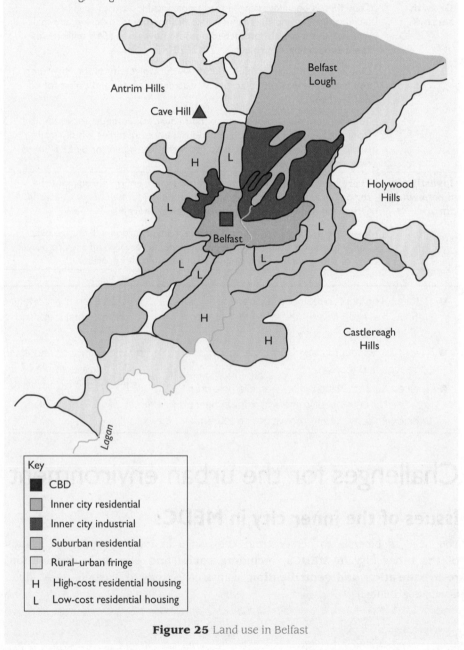

Figure 25 Land use in Belfast

Redevelopment is when an area is demolished and redesigned. In the inner city this might mean that a street of terraced houses is knocked down and replaced with a block of flats or other cheap housing.

Regeneration is when an area is upgraded. The aim is to improve the social and economic conditions. It happens in areas with issues of dereliction, pollution and out-migration. It might involve restoring old buildings and redesigning them for a different use.

Gentrification occurs when an area is demolished and upgraded, attracting richer people to live in the new, modern, expensive properties.

Social and economic deprivation

In the 1960s and 1970s, houses within inner city Belfast were in poor condition. Much of the heavy industry that had employed people was in decline. Factories were closing and work was increasingly scarce. Employment became less centralised and workers sometimes had to travel to the edge of the city. The inner city was in crisis, its identity as the centre of industry was under threat and it began to experience a spiral of decline. Many had to move out, while those who were left had less money to spend, and shops and services began to close down.

According to the most recent Northern Ireland *Multiple Deprivation Report*, some of the areas of highest relative deprivation in NI were found in inner city areas — the Falls, New Lodge, Shankill, Crumlin Road, Ardoyne, Upper Springfield and Whiterock.

This report was generated using 52 separate indicators of deprivation, which are divided up into: income (25%), employment (25%), health and disability (15%), education, skills and training (15%), proximity to services (10%), living environment (5%) and crime and disorder (5%).

Some of the most up-to-date deprivation figures illustrate the issues in the inner city.

Economic data

Unemployment: the current unemployment rate for NI (those of working age who are claiming unemployment) is 5.4%. The rate for Belfast is 7.4% (13,816 people). In north Belfast it reaches 16.9% in Duncairn and 13.5% in New Lodge while in west Belfast's Shankill it is at 13.5%.

Social data

Free school meals: one useful social measure is the extent of free school meals in an area. This is a guide to the number of families who are on low incomes. In 2012/13, 40% (5,333) of students attending secondary (non-grammar) schools received free school meals. This compared with 7.4% of grammar school students.

School attendance: another measure used is absenteeism (% of total half days) at secondary schools. Falls recorded 8.8%; Duncairn and Shankill were both at 11.2%.

Qualifications: a final measure of deprivation is the proportion of working-age adults with no qualifications. For example, in north Belfast's Water Works 32% of the population have no qualifications; in Duncairn the rate is 41%.

The people in these areas are in a cycle of poverty and struggling to break out of it. Disadvantaged children are more likely to fail at school, will leave with few qualifications and will be less likely to get a good job.

Re-urbanisation

Re-urbanisation is the movement of people back into an area that had previously been abandoned. The key feature of any re-urbanisation is that people from the suburbs or from outside the city make a decision to move back into the city. Usually the stimulus for any movement will come as a result of an intervention by the government to improve an area.

The development of Laganside (also an example of gentrification) and the Titanic Quarter are both attempts to re-urbanise areas of Belfast that experienced serious decline.

Knowledge check 26

What is the difference between redevelopment and gentrification?

Examiner tip

It is essential that you know facts and figures to show an understanding of the level of deprivation in the inner city.

Titanic Quarter

Much of the traditional manufacturing base in twentieth-century Belfast was located in Queens Island. The decline of Harland and Wolff from over 20,000 workers to 500 by 2002 meant that there was a large area of derelict land in the Belfast inner city (around 185 acres), with little opportunity for employment or economic growth.

Building started in 2006 with the aim of creating a new, fresh and modern space where people would be able to come to live, work and play. The hope was that new transport and communication links would allow this area to be the most accessible area in the city, and that this development would breathe new life into Belfast. Key aims included:

- the building of up to 5,000 dwellings
- the creation of a high-quality business area
- commercial development
- a Titanic signature project
- a major educational, third-level campus (Belfast Metropolitan College)
- restoration and conservation of the former Harland and Wolff headquarters
- the creation of hotels and other tourist accommodation
- the development of new leisure facilities, restaurants, cafes/bars and health clubs

In many ways, the first release of residential accommodation took place at the wrong time because the global financial crisis and recession took hold and much of the proposed development has had to be scaled down in recent years. However, the expansion of the area continues, with renewed plans for more accommodation to be built in the near future.

Gentrification

Gentrification (Table 9) is the development of run-down areas of a city so that they become fashionable places to live. Often the redevelopment means that the people who originally lived in these urban areas cannot afford to buy the new properties and are forced to move to other parts of the city.

Table 9 Features of gentrification

Positive features	Negative features
• Houses are improved — old, dilapidated buildings are regenerated • House values increase • New businesses to service the new, richer community are set up • Crime rates can fall	• A high demand for houses can cause problems • Increased house prices mean that original residents are stuck in the area or move out, and their children cannot afford to live in the same area • There can be conflict between the original and new residents

Laganside

The Laganside development started at the Lagan Weir, which is used to control the amount of water in the river upstream. The Laganside Corporation became responsible for developing 140 hectares of land alongside the River Lagan and 70 hectares in the Cathedral Quarter.

By the time the project was completed the Corporation estimated that over 14,200 permanent jobs had been created. A total investment of £939 million resulted in the creation of 213,000 m² of office space, 83,000 m² of retail and leisure space and 741 housing units.

Some houses in the Mays Meadow area and at Lands End Street, Laganview Street and Newfoundland Street (all around Bridge End) were removed and new apartment blocks were put in their place.

Many of the communities who lived beside the Laganside development felt that they had been ignored and felt no positive impact from the regeneration.

Examiner tip

Make sure that you understand the difference between re-urbanisation and gentrification, and be ready to give examples of different projects around Belfast to support your answer.

Issues of rapid urbanisation in LEDCs

You need to be able to demonstrate knowledge and understanding of **issues of rapid urbanisation in LEDCs, including informal settlements, service provision and economic activity**, with a case study to illustrate these issues (e.g. Nairobi, Kenya).

In the 1970s and 1980s many LEDC cities experienced a rapid expansion as many poor people left their small patches of land in the countryside and migrated into the urban areas. This, in turn, created huge problems for the cities, which had to grow to cope with all of the people.

Generally, when an LEDC expands like this, an economically segregated area develops on one side of the CBD, with expensive accommodation including high-rise apartments and some gated communities. Adjacent to this, and found on more marginal land, poor-quality housing develops into shanty towns or slums.

Wedges or ribbons of development encourage factories to be built — usually along the routes of major roads (towards ports or airports) or along railway lines. A surrounding ring of more mature suburbs/medium-cost housing is often found beyond this (Figure 26).

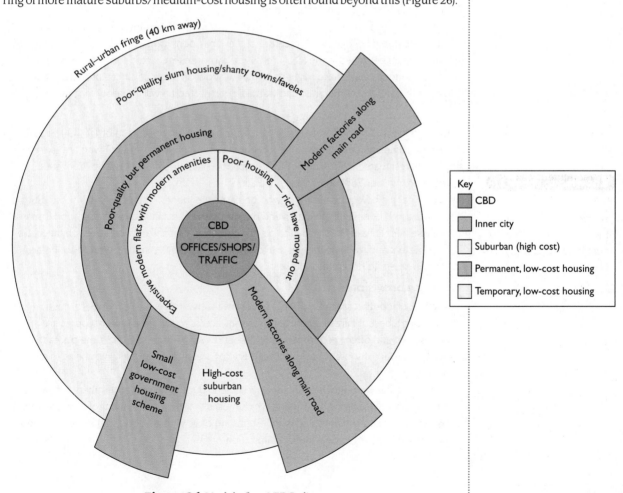

Figure 26 Model of an LEDC city

Unlike in MEDC cities, parcels of land on the outskirts (rural–urban fringe) are used to house the most affluent in society, but these are often interspersed with slum areas that develop near rubbish tips and poor-quality land.

The development of the LEDC cities in this way has led to a number of major problems.

Examiner tip

Make sure that you can describe the reasons for the rapid urbanisation of the area — what are the different push and pull factors at work on people in rural Kenya?

Case study

Nairobi, Kenya

Many African cities were originally built as colonial staging posts. Nairobi was set up in 1899 by the British colonial authorities as a railway camp on the Mombasa to Kisumu line. However, it developed quickly and by 1907 it was established as the capital of Kenya. The population of the city grew to 1.3 million in 1990, 2.1 million in 2000 and by 2009 was estimated at 3.1 million.

Informal settlements

One of the most common features of any large urban settlement in an LEDC is the growth of informal settlements — 'shanty towns' or slums. These areas are usually found in low-lying areas where there is a risk of flooding, with poor soils, close to rubbish tips, or on steep slopes where landslides might be common.

Few people in slums own the shacks they live in. The owners are usually the original builders of the dwellings, which means that they are poorly constructed with whatever materials can be gained in the local area. Typically houses are small (about 3m × 3m) with a floor made of mud. There might be a small window on more advanced properties but usually the construction of mud and wattle or corrugated tin will not allow for this.

Kibera slum

One infamous slum in Nairobi is Kibera. It has an estimated population of 170,000 people (though some sources estimate the population to be as high as 1 million) who are extremely poor. An estimated 2.5 million of the Nairobi population live in some sort of slum area, representing over 75% of the total population.

Few people have jobs and those who do earn little money. There are many social issues, such as a high level of HIV infection and regular cases of assault and rape. Education is sparse because people have little money to pay for schooling. Clean water and good sanitation are non-existent and life expectancies are low because people live in these squalid conditions with no access to decent healthcare.

Knowledge check 27

What are the main characteristics/features of the informal settlements/ slums in an LEDC city like Nairobi?

Service provision

The lack of jobs and space for people to live means that there is little money to invest in building a decent infrastructure and providing access to services. Water supplies are inconsistent and often require people to collect drinking water from the same places that are used for sewage disposal. Waste disposal services are entirely lacking, which helps to spread disease.

The land in Kibera is owned by the government, but the Kenyan government refuses to officially acknowledge the settlement, which means that the development of basic services like schools, clinics and toilets is limited. Shops are basic with any food sold being old, poorly stored and exposed to heat and contamination.

Until recently there was little access to water or electricity in Kibera. UN-Habitat has been involved in getting electricity to about 20% of the residents, but it can be expensive. Any medical care is provided by charities and religious groups — global charities have set up clinics where people can have a free HIV test and gain access to free ARV treatments if they are found to be HIV positive.

Economic activity

Jobs are hard to come by in the slum. Most of the people lack any form of education and are unskilled. There are few 'formal' jobs for people so the ever-increasing population has to resort to 'informal' service-related jobs in order to survive. These include washing car windows, picking out recycling materials at the city dumps and prostitution, and are carried out at great personal risk and often with very little reward.

Some transnational companies set up factories in LEDC cities to benefit from the cheap labour. The available workforce is so big that employers can demand difficult and binding conditions, with employees happy to work hard for little reward just so they can stay off the streets.

Summary

- In MEDC cities, the inner-city zones often present unique challenges in relation to social and economic deprivation.
- Poverty can have a big influence on lifestyles, education, job prospects and life expectancy.
- Inner-city areas require government intervention to get the area out of a 'spiral of decline', so re-urbanisation and gentrification projects are used (such as Laganside and the Titanic Quarter in Belfast).

- In LEDC cities, the challenges of urban living are caused by rapid urbanisation (and in-migration).
- This rapid urbanisation creates pressure on space, which leads to informal settlements/slums throughout the city and problems of service provision.
- The rapid urbanisation can create both opportunities and issues for the economy (e.g. Kibera in Nairobi, Kenya).

Topic 3 Development

The nature and measurement of development

The problems of defining development

There is no agreed definition for 'development'. When we look at the world we live in, we might use certain standards to indicate the advancement of a particular place. Most geographers accept that the study of development considers the quality of life for people living in a particular area. This will take into consideration wealth, aid, healthcare, education, poverty, infrastructure, political landscape, economics and environment.

The World Bank notes 298 separate indicators that can be used to analyse the differences between countries. This helps to highlight the division there is across the world. Over the last 200 years a 'development gap' has opened up between the rich MEDCs and the poorer LEDCs.

Examiner tip
Make sure that you have a good understanding of what development is, but also note that coming up with one comprehensive definition is difficult.

Economic measures of development

Economic measures of development help us to work out how much money or wealth there is within a country and how people actually earn that wealth. This is one of the easiest ways of measuring how people from one country compare with people from other countries.

Gross national product (per capita) or gross national income

GNP per capita or GNI measures the total economic value of all of the goods and services provided in a country through a year, divided by the number of people who live in that country. The amount is worked out in US dollars so that an easy comparison can be made. The higher the GNP, the more developed the country. At present the country with the highest GNI (with $87,030 per person) is Qatar, while the lowest is the Democratic Republic of the Congo, with a GNI of $350 per person.

Positive issues

- The comparison between countries is easy to understand.
- It gives a baseline of the amount of money in dollars per head of population that is earned within the country.
- It uses a common currency as a benchmark, which simplifies global comparisons.

Negative issues

- Working out how much money is earned within a country in a year is increasingly complicated to do — especially with the global markets.
- It can be affected by changes in currency rates — 1 dollar in one country might not buy the same amount of food as 1 dollar in another country — so the real purchasing power of money can vary from one place to another.
- It is a crude method of working out how much money is in a country's economy, and it fails to highlight that the distribution of this money within the country is uneven.

Percentage of people employed in primary activities

Sometimes we can learn how developed a country is by looking at the percentage of the workforce employed in the primary sector. A rich, more developed county is likely to have more people working in the secondary, tertiary and quaternary sectors. For example, the UK has around 2% of the population working in agriculture, while Vietnam has 73%.

Social measures of development

Social measures of development are used to assess how well a country is developing in areas that affect people directly, for example healthcare, education and diet. These measures help to indicate the quality of life of individual people in a country.

Healthcare

Life expectancy is a good example of a measure of healthcare. It measures the average lifespan that someone born in a country can expect. It can be affected by wars, disease and natural disasters, but can show how developed the medical facilities in a country are. The general rule is that the higher the life expectancy, the more developed a country is. People in the UK have a life expectancy of 78 years, whereas Ugandans have a life expectancy of only 41 years.

Infant mortality rate measures the number of children who die before they reach the age of 1 (from every 1,000 live births, per year). This helps to determine the quality of ante- and post-natal services in a country. Generally, less developed countries spend less money in tackling issues affecting infant mortality, while more developed countries spend more and have a much lower rate. In the UK this figure is around 6 per 1,000 live births per year, whereas in Sierra Leone the rate is 195.

Positive issues

- Life expectancy and infant mortality rate are easy to work out. They depend on good vital registration data, which can be found in most countries.
- They both give a useful figure that allows direct comparison with other countries and inferences to be made about the available health services.

Negative issues

- Not all countries have robust methods of recording these vital registration data and in very poor countries information might be inaccurate.
- Some aid organisations argue that variation in the amount of money allocated to tackling particular health issues might not give a true picture of the overall state of healthcare in a country. For example, political decisions might have been taken to not spend money on ante-natal care.

Education

One social measure of education is the adult literacy rate. This is the percentage of the adult population who are able to read and write. In the UK, and most other MEDCs, over 99% of adults have learned to read and write, yet in Somalia only around 24% of adults have achieved this.

Diet

Increasingly, international agencies such as the United Nations (UN) and the World Health Organization (WHO) have been using calorie intake to measure variation in diet across the world. In wealthy countries like the USA the average calorie intake per person is 3,725 kilocalories per day, which contrasts with poorer nations like Eritrea (1,555 kilocalories).

Composite measures of development

The human development index

In response to an over-reliance on simple measures of development, the UN looked for a measure that would combine some of the major indicators into one easy-to-use measure. The human development index (HDI) is worked out using three measures: life expectancy (a social measure), education (the mean number of years of schooling — a social measure) and gross national income per capita (an economic measure).

Examiner tip
There are many different social issues that can be used as measures of development — healthcare is an important one to evaluate, but make sure that you know at least one more social measure in detail.

Knowledge check 30
How does life expectancy help us understand healthcare as a social measure of development?

Each of the three measures that make up the HDI is ranked so that a high performance will be given an index mark approaching 1 and a poor performance will be towards 0. Each year the UN Development Programme (UNDP) updates its *Human Development Report*, which ranks countries across the world (Tables 10 and 11).

Table 10 Top five countries (very high human development index, 2011)

Country in rank order	HDI value	Life expectancy (years)	Mean years of schooling (years)	Expected years of schooling (years)	GNI per capita ($PPP)
1 Norway	0.943	81.1	12.6	17.3	47,557
2 Australia	0.929	81.9	12.0	18.0	34,431
3 Netherlands	0.910	80.7	11.6	16.8	36,402
4 USA	0.910	78.5	12.4	16.0	43,017
5 New Zealand	0.908	80.7	12.5	18.0	23,737

Table 11 Bottom five countries (very low human development, 2011)

Country in rank order	HDI value	Life expectancy (years)	Mean years of schooling (years)	Expected years of schooling (years)	GNI per capita ($PPP)
183 Chad	0.328	49.6	1.5	7.2	1,105
184 Mozambique	0.322	50.2	1.2	9.2	898
185 Burundi	0.316	50.4	2.7	10.5	368
186 Niger	0.295	54.7	1.4	4.9	641
187 Congo (DR)	0.286	48.4	3.5	8.2	280

GNI per capita ($PPP) is the GNI per capita based on purchasing power parity (PPP). This converts the usual GNI into international dollars so that an international dollar has the same purchasing power as a US dollar in the USA.

Positive issues

- HDI takes into account both social and economic data and the social data include figures based on health and education, making this a more rounded and realistic measure compared with merely social or economic indicators.
- The information is updated annually, which allows countries to monitor their development progress and smooth out any fluctuations from one year to the next.

Negative issues

- Some argue that wealth has too much importance within the HDI and that this can adversely influence the ranking of a country within the table.
- Some suggest that the HDI takes too simplified an approach to measuring development and note that other factors beyond education, life expectancy and GNI are needed to get a more rounded picture of what it is like to live in a particular country (for example by referring to markets, industry or environmental issues).

Examiner tip
The HDI composite measure is a popular one in exam papers so make sure that you are well practised in being able to describe and evaluate it. However, questions might also test your knowledge of a second composite measure, so be ready!

Knowledge check 31
What are the three main indicators used in the HDI?

The physical quality of life index

This is another composite measure that uses life expectancy at year 1, infant mortality rate and basic literacy rate — each having an equal weighting. The values range from 0 to 100 and countries are ranked accordingly.

This was originally developed as an alternative to GNP, but the UN HDI became more widely used. It was often criticised because the three strands of the measure are similar and depend on each other: a low life expectancy indicates that money is not spent on healthcare, which will cause a high infant mortality rate. A low literacy rate indicates that money is also not spent on education of young people.

Knowledge check 32

Why is the HDI a better measure than the PQLI?

Regional contrasts in development

You need to be able to identify and explain regional contrasts in development, including a national-scale case study to demonstrate distinct regional variations in development.

Different rates of development can occur within countries, with some areas seen as being more important and requiring more investment. This has led to the core–periphery model, whereby economic growth or development takes place more quickly in a core region, while other more peripheral areas, which might be further away from the source of power, are ignored and receive little investment.

The core is most likely to include the capital city and the country's main industrial areas. These areas provide a large economic base and market, which create the impetus for further growth and development. Resources and services are generally better in the core than in other areas, although much of this is down to there being a large population that can sustain the services.

The periphery areas are generally far from the capital city, where the majority of political decisions are made. Resources and services are noticeably poorer and there are fewer opportunities and jobs available. This can lead to a tradition of migration from the peripheral regions towards the core.

Reasons for this disparity between the two regions include the following:
- **Physical reasons:** core regions are often located around major rivers on fertile, flat land, which is easy to build on. Many peripheral regions are separated by aspects of physical geography — mountains, poor soils and extreme climates make it difficult to establish industry and services, making development less likely. Areas like Northern Ireland, which are separated from the mainland by sea, are also less likely to be developed.
- **Economic reasons:** often industry arises where valuable raw materials are found — for example, many urban areas developed in the Welsh valleys as coal and iron ore were found. Investment will generally fund the exploitation of new resources until they are exhausted. However, this investment is usually withdrawn when these resources run out. As commerce develops in major cities this can continue to build the importance and reputation of the core.
- **Historical reasons:** often the reason for the initial establishment and development of a place is lost through time but its importance is maintained as it is set aside as the capital city or seen as a regional capital. Edinburgh, for example, has long been

the second biggest city in Scotland but retains its importance and status, with the Scottish parliament being established here.

- **Political reasons:** governments often have to make decisions and take action to protect or establish the development of a place. This can involve the setting up of regional development agencies tasked with trying to improve and attract investment into an area, while other areas are overlooked.

Case study

Regional variations in the development of Italy

Few countries in Europe display a bigger regional variation than Italy. The differences between the north and south of the country have increased with the development of the country over the last 50 years. Today the OECD recognises that there are huge socio-economic disparities between the wealthy and developed north compared with the Mezzogiorno region of the south.

Northern Italy

- **Physical geography:** the area is dominated by flat, fertile land (especially along the River Po). Rainfall is adequate for farming (800 mm) and temperatures provide a long growing season. There is much successful farming, for example olives and wine growing in Tuscany.
- **Economic development:** the main reason for the growth and development of the north is the accessibility to the rest of western Europe. Many of the cities, like Turin, Milan and Genoa, grew up as industrial centres following the Second World War. Car manufacturers like Fiat and Maserati built many factories and the components needed were often manufactured in surrounding factories (for example, Pirelli tyres). The GNP increased at a rate of between 6% and 10% per year from the 1960s to the 1980s. As more money came into the area, this allowed services and retail businesses to grow and the economy continued to take off.
- **Political factors:** the capital city Rome sits at the southern edge of this region. Rome is the core region in Italy — at the centre of economic and political activity as well as power.

Southern Italy

- **Physical geography:** the area is dominated by the Appenine Mountains, which run up the spine of Italy. Soils are generally shallow and not fertile. There can be drought conditions for much of the year as rain is scarce (often less than 500 mm per year). Farming is a struggle and while many people continue to attempt to farm this area, the returns are not great and farmers live a subsistence existence.
- **Economic development:** the Mezzogiorno region is far from the central areas of Europe. There has been little economic development because of the costs of transporting goods to the north and into the rest of the European market. As a result the area has suffered from mass out-migration — many young people in particular have travelled north in search of better jobs and opportunities. This has accentuated the problems and led to economic decline in the area.

The main source of income is farming, but this is limited by the climate (see above). Deforestation and soil erosion are ever-increasing threats. The tradition of sub-dividing land between children has also meant that farms are now too small to support family groups.

- **Political factors:** the area has traditionally been seen as a political desert, with a history of being attacked by various colonial rulers. Crime and corruption from the Mafia have had further negative impacts.
- **Government and EU intervention:** there has been a long history of interventions aiming to address the imbalance between the north and south, including the 'Cassa per il Mezzogiorno'. Since 2007 the main policy has been the National Strategic Framework, which aims to promote the under-used potential of different areas by funding infrastructure projects and encouraging entrepreneurial projects.

Knowledge check 33

What are the main reasons why the south of Italy is a lot more deprived than the north?

Summary

- The concept of development is not easily defined but it usually relates to the quality of life that people experience in an area.
- Two key economic measures of development are gross national product (per capita) or gross national income and the percentage of people employed in primary activities.
- Two key social measures of development are healthcare (including life expectancy and infant mortality rate) and education (literacy rate).
- Two key composite measures of development are the HDI (human development index) and the PQLI (physical quality of life index).
- Regional contrasts exist between and within countries.
- The core–periphery model helps us to understand that core regions (like the capital city and major industrial areas) get more investment than peripheral areas (far away from the core regions).
- These regional contrasts are controlled by physical, economic, historical and political factors.
- In Italy the north is the core region and is more developed. The south is the peripheral region and is less developed.

Issues of development

Colonialism

Colonialism has had a big impact on the development of many European countries across the globe, and continues to do so today. From the fifteenth century, European countries sent various expeditions out around the world to explore and conquer new territory. Spain and Portugal claimed territory in Central and South America. France concentrated on the west of Africa and Britain extended an empire that at one point covered over a quarter of the Earth's land surface (Figure 27).

Being part of a wider European colonial power brought advantages and disadvantages:

- **Advantages:** the European countries spent money investing in the infrastructure of their conquered lands: roads, railways, hospitals, schools and government buildings were all constructed in the style of the home nation. These often helped to improve the standards of living in the country. Exotic foods and materials were exported back to the home country, diversifying diets and fashions.
- **Disadvantages:** the European countries often ruled by force and were ruthless in imposing decisions on their colonies. They forced people to fight in their armies (through the two World Wars), taxed them and took many of their valuable assets. For example, much of the mineral wealth of colonies in South America and Africa was stripped and taken back to Europe.

Examiner tip

It is easy for students who are interested in history to go into too much depth in questions about colonial empires. Make sure that your answer addresses the question set, which will often be about the impact of colonialism.

Knowledge check 34

Is colonisation usually a good or bad thing for a country/area?

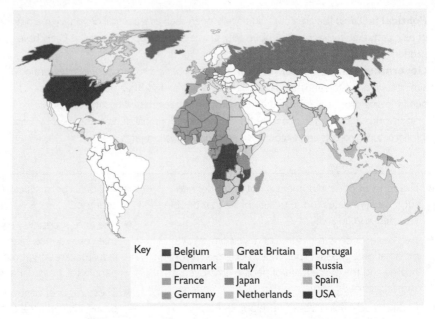

Key
- Belgium
- Denmark
- France
- Germany
- Great Britain
- Italy
- Japan
- Netherlands
- Portugal
- Russia
- Spain
- USA

Figure 27 The colonial possession of the world in 1914

Neo-colonialism (dependency)

Following the Second World War, many western European countries did not have enough money to sustain a global presence and they started to give independence back to the colonies. In some cases, newly independent countries joined informal groups that helped to maintain the historic link to a 'mother country', for example the British Commonwealth.

In the twentieth and twenty-first centuries a new type of colonialism has started to dominate many LEDCs. Economic control through transnational corporations has led to a new dependency relationship in which poor countries make big deals with companies that invest in the country and provide jobs for the people. In return the companies are involved in political decisions, awarded governmental contracts and given preferential treatment. For example, many Chinese companies have built roads in Africa in exchange for mining rights to exploit natural resources.

The main dependency relationship is linked to debts and repayments of loans that were made in the past. European banks offered large loans so that newly independent countries could improve their infrastructure, but this often led to big debts that never seemed to reduce over time. This has allowed some MEDCs to continue to dictate control over their ex-colonies.

Globalisation

Globalisation refers to the world becoming more interconnected and interdependent. People around the world have become a lot more connected through developments in technology, communications and the internet. Globalisation is also evident in trade

Examiner tip

This is often a difficult concept to explain in examination answers so make sure that you are well prepared to discuss the recent linkages formed between countries and companies.

arrangements as goods and services can now be easily and quickly moved from one part of the world to another.

Many transnational companies have factories and offices in countries all over the world — for example, Ford Motors and Cadbury.

Some of the key features of globalisation are as follows:
- Globalisation has brought the world's economies closer together (especially in relation to trade and investment).
- Trade across the world has grown quickly, with many LEDCs increasing their industrial output.
- Global communications have been a key factor in the continued growth of links across the world.
- Profits made by multinational companies (MNCs) often flow back to the head offices in MEDCs and funds rarely remain in the LEDCs. However, this is changing as new financial and tax rules are introduced.
- Individual countries are less independent than they used to be. Global firms like Exxon, Shell and Walmart can have more money and more control over decisions than some governments.

Advantages of globalisation

- Globalisation is responsible for many formal employment opportunities in LEDCs.
- Often MNCs will spend a lot of money helping to improve the infrastructure and the social conditions of an area. LEDC governments will go out of their way to attract investment from big firms.
- Additional factories mean that more money is flowing into and through the country.
- Workers can receive a better education and are able to improve their skills, which makes them more employable and able to earn more money in the future.
- New skills, techniques and technology are brought into poor countries, allowing them to develop.

Disadvantages of globalisation

- Although many LEDCs embrace any jobs that are brought into their country, these are often less well paid than similar jobs in MEDCs. This pay disparity makes LEDC workers feel less valued than those in MEDCs.
- Working conditions are not always as good as they are in MEDCs. Many factories are not owned by the MNC but are contracted out so that usual working standards do not have to be implemented and corners can be cut on hours and conditions.
- As most of the profits for anything manufactured and sold flow straight back to the headquarters of the company, little of the money stays or is invested in the LEDC partner.
- There is little job security. MNCs can be ruthless operators and will only work in partnership with a country as long as it is economically viable and safe to do so. At the first sign of difficulty a company might pull out and leave workers without a job or income.

Knowledge check 35

What is a definition of globalisation?

Aid

Aid involves one country or organisation giving resources to another country. This might be in the form of money, expertise (such as aid workers, healthcare professionals or rescue workers) or goods (water, food, blankets, tents, shelters, tools, rescue equipment and transport facilities).

Different types of aid

- Bilateral aid is when help is given from one country to another country. The aid is usually tied, so that the donating MEDC can direct the money towards particular issues and priorities.
- Multilateral aid is when aid comes from world/international organisations like the World Health Organization (WHO) and the United Nations (UN). Money is paid into the organisations and then used to fund projects and rescue missions all round the world.
- Voluntary aid is when charities and non-governmental organisations (NGOs) are set up to help for a particular issue or to support a particular place. These are funded by the public. Charities like Christian Aid, Comic Relief, Oxfam and Fields of Life often provide help and workers on short- and long-term development projects.

Knowledge check 36

What is the difference between bilateral and multilateral aid?

Advantages of aid

- Aid can make a huge difference to the lives of people in LEDCs.
- Short-term and humanitarian aid can help save lives in emergency situations.
- Charities can make a big difference by working in small communities and working alongside partner organisations in LEDCs.
- Aid can help to improve the standard of living of ordinary people in LEDCs.

Disadvantages of aid

- Aid does not always make a difference to the communities that it is aimed at.
- Sometimes it does not reach the people who need it most because of corruption and poor administration systems.
- Many LEDCs have become reliant on a regular flow of aid. They become less able to stand on their own two feet without support.
- Tied aid can put strict conditions on LEDCs, which makes the aid less desirable.
- Aid can cause problems for local producers because any food aid that comes into the area might bring the food prices down and undermine any profit that local farmers might make.

Examiner tip

Make sure that you have learnt the different definitions of aid and see if you can quote examples of where they have been applied.

Trade

Trade is the flow of goods and services between producers and consumers around the world. Good trade links are necessary for any country to be able to make money.

Each country achieves a balance of trade (the difference between the total cost of all imports and the value of its exports). As a result, some countries have a trade surplus — they export more than they import — and others have a trade deficit.

LEDCs make much of their money through their exportation of primary products, which they export in a raw state to the MEDCs for processing. For this reason they are affected by price fluctuations on the raw materials. Many LEDCs do not have the technology to process the raw materials. In some cases this is because MEDCs put tariffs and charges on processed goods, thus discouraging LEDCs from carrying it out.

However, in recent years globalisation has led to LEDCs taking a bigger share of the global manufacturing industry. Electronics and textiles industries have been growing rapidly within LEDCs. A group of newly industrialising countries (or NICs) have expanded their manufacturing. China, for example, now has a 17% stake in all African exports, with Brazil and India accounting for another 9%.

Debt

Without taking out large loans, many LEDCs cannot build the infrastructure that they need to expand and grow. They turn to richer countries, the World Bank and the International Monetary Fund for loans for specific projects such as infrastructure, hospitals and schools. This money needs to be repaid with interest and sometimes this can take many decades.

Money that is tied up in paying the interest payments from a debt cannot be used to fund further development projects. Many countries fall into the trap of borrowing too much money and suffer further as the lending countries add additional clauses to control rates of interest or tariffs.

The Jubilee 2000 organisation noted the following:
- Each year developing countries pay over nine times more in debt repayments than they actually get in grants.
- The world's poorest countries pay over $23 million every day to the rich world.
- Countries across the whole continent of Africa spend four times as much making the interest payments on debts as they do on healthcare.

Suggested solutions to the debt issues faced by countries are as follows:
- Rescheduling — capital and repayments can be rescheduled to reduce their impact on the country.
- Debt-for-nature swaps — some international bodies (like the World Wildlife Fund) have lobbied for countries to receive debt relief and cancellation when they put measures in place to preserve an area of global importance (like the rainforest).
- Debt-for-equity swaps — sometimes unpaid debt will be invested back into schemes for future development projects within a country rather than making money for a foreign bank or country. Unpaid debt is exchanged for equity or property or for land/mining rights.
- Debt cancellation — organisations like the Jubilee Debt Campaign argue that debts for many LEDCs should be written off to allow these countries to stand on their own two feet.

Examiner tip
Take a look at some of the issues that might still limit trade, such as trading blocks and tariffs, and take notes on these.

Knowledge check 37
What is the difference between a trade deficit and a trade surplus?

Examiner tip
The issue of debt relief is a contentious one. Investigate this further and consider whether you think debt relief is actually of benefit to an LEDC.

Knowledge check 38
What are the disadvantages of debt relief for MEDCs?

Case study

Issues that affect, positively and negatively, the development of one LEDC (Ghana)

The Republic of Ghana is located on the west coast of Africa, on the Gulf of Guinea. It is an LEDC but recently its economy has been growing and it is developing at a fast rate. Ghana is a producer of petroleum and natural gas. It is also one of the largest gold and diamond producers, as well as the world's second largest producer of cocoa.

Colonialism

Contact was first made with Europeans in the fifteenth century when Portuguese traders established the Portuguese Gold Coast in west Africa. The Dutch, Danish, Germans and Swedish also set up gold trading ventures along the coastline — building forts and castles. In 1874 the British established control over the area and it was called the British Gold Coast. Accra was established on the coast as the capital city. The British invested money in schools, the road and rail network, and communications.

Probably the biggest impact was the introduction of the cacao tree in 1878 as a cash crop. Within 50 years Ghana was one of the world's leading suppliers of cocoa.

Ghana fought long for self-governance and became the first sub-Saharan country to gain independence from the British in 1957. However, the political journey since then has not always been easy.

Neo-colonialism (dependency)

The history of Ghana is steeped in its association with a wealth of natural resources — minerals and oil. In recent years the country has developed links with MNCs and with countries like China in order to continue developing trade and exports. New digital-based manufacturing plants have been built to produce tablet computers and smart phones.

Multinational companies have been involved in stimulating Ghana's economy and creating new jobs. These include mining, petroleum, telecommunications and banking companies such as MTN, Vodafone, Guinness, Unilever and Zain. Ghana has even set up its own telecommunications MNC called Rig Communications.

One major industry that has links to many different MNCs is cocoa farming. Ghana has around 720,000 cocoa farmers and the production of cocoa is a major source of revenue. In 2008, Cadbury set up the Cadbury Cocoa partnership, aiming to invest £45 million over 10 years to help Ghana's cocoa farmers.

Globalisation

Ghana's Takoradi Harbour was built in 1928 and encouraged the country to develop one of the world's fastest-growing shipping industries.

In recent years the GDP in Ghana has increased dramatically — by 14% in 2011 and 8.7% in 2012. Much of this is down to the global trading and business links that Ghana has established with other countries around the world.

Ghana recently adopted an economic plan called 'Ghana Vision 2020', which aims to eventually get Ghana accepted as an MEDC. Before this can happen it needs to be accepted as a newly industrialised country.

Aid

Even though the economy in Ghana is growing faster than that of western Europe, it still relies heavily on aid. The country receives $1 billion a year in aid, which makes up around 10% of the country's GDP.

One controversial aid project was the construction of the Akosombo Dam on the river Volta between 1961 and 1965 to produce cheap electricity, which resulted in the relocation of 80,000 people from 700 villages. The dam was built using bilateral aid from the UK and USA and multilateral aid from the World Bank. American firms were brought in to complete the work.

Trade

The economy of Ghana has long been founded on trade. Cocoa, gold, diamonds and timber are all major commodities that are exported around the world. Oil and natural gas are extracted and exported overseas. Unfortunately, the narcotics trade has also taken root in Ghana in recent years.

Debt

At the 2005 Gleneagles Summit, Ghana was allocated $4.2 billion of multilateral debt relief (out of the country's $6 billion debt). This is freeing up about $156 million each year, to be spent on other projects.

Examiner tip

This case study is a starting point for looking at how development issues might impact one country. In an exam you need to build on this with more facts and figures to support your answer, and show a thorough understanding of how each concept works.

Knowledge check 39

Describe the importance of Ghana's location in its development.

Summary

- Development is affected by a number of issues that can slow down or speed up the process.
- Colonialism had a big impact on many countries in the past and brought some stability and investment.
- Neo-colonialism (dependency) is replacing colonial relationships as countries make deals with MNCs.
- Globalisation is a major influence on the global economy, bringing both benefits and challenges.
- Through globalisation, countries are becoming more connected and more interdependent.
- There are many different types of aid, including bilateral, multilateral, voluntary and tied.
- Aid is not always a good thing — countries can become too reliant on it.
- Trade is the flow of goods and services between producers and consumers around the world.
- Debt is a big problem for LEDCs. Loans taken many years ago are still being paid off.
- The case study of Ghana illustrates the main issues that affect development: colonialism, neo-colonialism (dependency), globalisation, aid, trade and debt.

Questions & Answers

In each AS Unit 2 Geography paper there are seven questions:

	Compulsory?	Marks (out of 90)	Exam timing (out of 90 minutes)
Section A			
Q1 Skills and techniques	Yes	30	30
Section B			
Q2 Population: short questions	Yes	12	12
Q3 Settlement and urbanisation: short questions	Yes	12	12
Q4 Development: short questions	Yes	12	12
Section C			
Q5 Population: essay question	Answer two from questions 5, 6 or 7	24 marks = 12 marks for each question	24 minutes = 12 minutes for each question
Q6 Settlement and urbanisation: essay question			
Q7 Development: essay question			

Examination skills

As with all A-level exams there is little room for error if you want to get the best grade. Gaining a grade A is not easy in AS geography so you need to ensure that every mark counts. The following table shows the minimum UMS (uniform mark scale) that you need to access particular grades.

Grade	AS Unit 1 (out of 100)	AS Unit 2 (out of 100)	Overall AS marks
A	80	80	160
B	70	70	140
C	60	60	120
D	50	50	100
E	40	40	80

Each of the two AS exam papers is 1½ hours. There are 90 marks available on each, which means that you get 1 mark per minute to work your way through the paper. The main reason why so many students struggle with this paper is that they fail to manage their time appropriately and as a consequence they do not have enough time left to answer the essays at the end in sufficient detail. If you find that you have time left over in this exam, the chances are that you have done something wrong.

Exam technique

Students often find it difficult to break an exam question down into its component parts. On CCEA exam papers, the questions are often long and difficult to understand, so you need to work out what the question is asking before you move forward.

Command words

To break down the question properly, get into the habit of reading the question at least *three* times. When you do this it is sometimes a good idea to put a circle round any command or key words that are being used in the question.

A common mistake is failing to understand the task being set by the question. There is a huge difference between an answer asking for a discussion and one asking for an evaluation.

The main command words used in the exam are as follows:
- **Compare** — what are the main differences and similarities?
- **Contrast** — what are the main differences?
- **Define** — state the meaning (definition) of the term.
- **Describe** — use details to show the shape/pattern of a resource. What does it look like? What are the highs, lows and averages?
- **Discuss** — describe and explain. Argue a particular point and perhaps put both sides of this argument (agree and disagree).
- **Explain** — give reasons why a pattern/feature exists, using geographical knowledge.
- **Evaluate** — look at the positive and negative points of a particular strategy or theory.
- **Identify** — choose or select.

Structure your answer carefully

Sometimes the longer questions on exam papers — for example, question parts for up to 6 marks or essay questions for up to 12 marks — can be an obstacle for students. Later in this section we will look at some questions and give more guidance on how you should structure your answers.

One simple approach to consider is drawing up a brief plan for your answer so that you know where it is going and how you will cover all of the main aspects of the question. For example, you could draw a box to illustrate each element needed within an answer and fill each one with facts and figures to support the answer, using the marking guidance to help you work out how much time to spend on each section (examples are shown for the essay questions below).

Show your depth of knowledge of a particular place/case study

The essay questions on the exam paper are usually focused on giving the student the opportunity to apply knowledge and understanding of case study material to a particular question. It is really important to show what you know here.

Examiners are looking for specific and appropriate details, facts and figures to support your case. The better you know and understand your case studies, the higher the marks you can potentially achieve.

About this section

A practice test paper with exemplar answers is provided. This will help you to understand how to construct your answers in order to achieve the highest possible marks.

In order to cover the range of potential questions in the 'techniques' section, two practice questions are included.

Examiner comments

Some questions are followed by brief guidance on how to approach the question (shown by the icon ⓔ). Student responses are followed by examiner's comments. These are preceded by the icon ⓔ and indicate where credit is due. In the weaker answers, they also point out areas for improvement, specific problems, and common errors such as lack of clarity, weak or non-existent development, irrelevance, misinterpretation of the question and mistaken meanings of terms.

Question 1A **Geographical techniques**

Study the table below, which relates to an investigation into the relationship between **GDP** per capita and literacy rate across 15 countries.

Country	GDP per capita/ US dollars	Rank	Literacy rate/%	Rank	Difference in ranks (d)	d^2
Kuwait	45,455	1	94	5	−4	16
Ireland	43,592	2	99	2.5	−0.5	0.25
Finland	38,655	3	100	1	2	4
Saudi Arabia	31,729	4	86	9		
Argentina	12,034	5	98	4	−1	1
Brazil	11,909	6	90	6	0	0
Egypt	6,724	7	72	12	−5	25
Guatemala	5,102	8	69	13	−5	25
Samoa	4,517	9	99	2.5		
India	3,876	10	74	11	−1	1
Bangladesh	1,883	11	79	7	4	16
Kenya	1,766	12	87	8	4	16
Zambia	1,712	13	80	10	3	9
Burkina Faso	1,513	14	22	15	−1	1
Ethiopia	1,139	15	39	14	1	1
						$\Sigma d^2 = 182.5$

(a) **Complete the table by filling in the four missing values.** (2 marks)

Student answer to question 1A

(a)

−5	25
6.5	42.25

ⓔ **2/2 marks awarded** All the right answers are inserted in the correct place.

ⓔ **This is a straightforward task — 2 marks if all four are correct, 1 mark for two or three correct. Some students forgot to put the minus sign in front of the 5 and lost a mark.**

(b) **Calculate the Spearman's rank correlation for the data shown and make a comment on the statistical significance of the result. (The formula and significance graph and table are found on pp. 17–18.)** (6 marks)

(b) $r_s = 1 - \left(\dfrac{6\sum d^2}{n^3 - n} \right)$

$r_s = 1 - \left(\dfrac{6 \times 182.5}{15^3 - 15} \right)$

$r_s = 1 - \left(\dfrac{6 \times 182.5}{3,375 - 15} \right)$

$r_s = 1 - \left(\dfrac{1,095}{3,360} \right)$

$r_s = 1 - 0.33$

$r_s = 0.67$

The result is very significant and shows a strong positive correlation.

ⓔ 5/6 marks awarded The processing of the Spearman's rank correlation is carried out correctly but the answer does not go into enough depth when describing the statistical significance. You need to use the graph/table to work out that the answer is within the 95–99% arc. The question does not require comment on the correlation/relationship for the graph, so no marks are awarded for this.

ⓔ There are 2 marks for getting the first breakdown of the formula correct and another mark for taking the formula through before the division. There is 1 mark for getting the right r_s value (to 2 significant figures). The final 2 marks are given for the comment on the significance. 1 mark is awarded for noting that this is a significant result and 1 mark for noting that this is found between the 95% and 99% levels.

(c) Explain some geographical reasons that could account for this particular statistical result. (4 marks)

(c) The table and the Spearman's rank result show that the result is a positive correlation. This means that there is a link or a relationship between the two variables. As GDP increases this might have a positive impact on literacy. This is not surprising as governments in poor countries might not have much money to invest in the education system and this will have a big impact on literacy rates.

ⓔ 3/4 marks awarded This identifies the main relationship and goes on to explain a potential reason as to why this might be happening in this particular case. It could have gone into a little more detail to explain why this might be the case.

ⓔ As the GDP per capita increases there is a strong positive correlation with the literacy rate. This calculation shows that there is a strong relationship between the two variables. The answer needs to explain this relationship — maybe countries that are richer will have more money available to provide better/free/accessible education for their citizens. 4 marks are awarded for a well-argued response. Statistical reasons are not credited.

(d) Some geography students carried out a piece of fieldwork on the sand dunes at Portstewart Strand. They measured the amount of time that it took for 200 ml of water to completely infiltrate into the sand/soil across the sand dune. Their results are shown in the following table.

Distance from sea/m	0 (sea)	30	60	90	120	150	180
Infiltration rate/s	12	34	67	105	65	204	204

State the mode for the data in the table above. (1 mark)

(d) The mode is the most frequently occurring number in the data set, so 204 seconds.

ⓔ 1/1 mark awarded

(e) Calculate the median for these data and state *one* disadvantage of using this statistic to summarise the data. (3 marks)

(e) The median is 65. The median is not as reliable as other measures of standard deviation such as the mean because the median is not good when comparing small amounts of data, and it is difficult to use the median when going onto further statistical techniques.

ⓔ 2/3 marks awarded This shows good understanding and depth when discussing the problems/disadvantages of using the median in statistical analysis. However, it identifies the wrong value for the median so fails to score the mark.

ⓔ The median is the central value when all of the values are ranked in order. In this case the median is 67 (the fourth number in the sequence).

(f) Study Figure 10 on page 14, which shows the percentage of total electricity production by generating source for selected countries.
     The table below has been partially completed, illustrating the percentage contribution of hydroelectric, nuclear and thermal (and other) types of energy production. Use the triangular graph to complete the table. (4 marks)

Country	France	Sweden	USA	Canada
Thermal and other (%)	24			29
Hydroelectric (%)	25			59
Nuclear (%)	52			12

(f)

Sweden	USA
22	74
48	12
30	14

(e) **4/4 marks awarded** All the values are correct.

(e) **This is fairly straightforward. 1 or 2 correct answers = 1 mark, 3 or 4 correct answers = 2 marks, 5 correct answers = 3 marks. Answers must be ±1% to be awarded a mark.**

(g) When might a triangular graph be an appropriate method of presentation? (2 marks)

(g) A triangular graph is appropriate when you are investigating something that includes three parts, something simple and something that includes percentages.

(e) **2/2 marks awarded** This identifies both the need for three parts and the percentages, although understanding appears to be weak.

(e) **Triangular graphs can only be used in geographical studies when there are three separate variables and each component must be able to be measured out of 100%. There is 1 mark for identifying that only three variables are found and 1 mark for noting that we use percentages for this.**

(h) Study Figure 3 on page 9, which shows the world population density by country. Answer the questions that follow.
(i) Identify the population density for the UK. (1 mark)
(ii) Name the mapping technique used to show information like this. (1 mark)
(iii) Discuss *one* advantage and *one* disadvantage of using this type of map to present geographical information. (6 marks)

(h) (i) 200–500 people per km^2

(e) **1/1 mark awarded** There is 1 mark available for the right answer but the unit of measurement must be given.

(ii) Choropleth

(e) **1/1 mark awarded** 'Area-shaded map' is also acceptable.

(iii) Choropleth maps are easy to draw and show data in a simple and straightforward manner. However, it is argued that they can oversimplify the patterns in areas that are densely populated, and will not actually show the variation within the area or country.

ⓔ 4/6 marks awarded This goes into some depth in relation to both the advantages and disadvantages. There is enough for 2 marks out of 3 for the first point. The disadvantage is well stated and the variety within the shaded area is noted. This point is maybe even better than the first point but it still lacks clarity and therefore only gets 2 out of the 3 marks.

ⓔ 3 marks are available for a well-developed discussion of an advantage and 3 marks for a well-developed discussion of a disadvantage. Advantages might include that choropleth maps can be easy to read and can provide a good visual presentation of the data. Limitations might include the fact that sometimes the maps can be oversimplified so that large regions will only indicate one value, when the range within the area might be wide. Also, maps can sometimes provide big contrasts at borders that are unrepresentative of reality.

Question 1B **Geographical techniques**

A geography student used nearest neighbour analysis to investigate the distribution of Met Office automatic synoptic and climate stations across Northern Ireland in 2012. The following hypothesis was proposed:

The distribution of automatic weather stations across Northern Ireland is significantly random.

Study Figure 13 on page 20, which shows a map of the Met Office automatic synoptic and climate stations in Northern Ireland in 2012. The table below is a partially completed version of the nearest neighbour analysis of the distribution.

Met Office station	Number	Nearest neighbour	Distance/km
Giant's Causeway	1		20
Ballypatrick Forest	2	1	20
Portglenone	3	16	20
Killylane	4	6	12
Orlock Head	5	9	10
Aldergrove	6	4	12
Helens Bay	7		10
Stormont Castle	8	7	10
Ballywatticock	9	5	10
Murlough	10	12	20
Killowen	11	12	22
Katesbridge	12	10	20
Glenanne	13	12	25
Derrylin	14	15	22
St Angelo	15	14	22
Lough Fea	16	3	20
Castlederg	17	15	25
Banagher	18	19	20
Ballykelly	19		12
Magilligan	20	19	12

map area = 165 km × 130 km = 22,275 km²

$\sum d = 344$

(a) Using Figure 13 on page 20, complete the table by filling in the missing values.   (3 marks)

(a) Giant's Causeway, 2; Helens Bay, 9; Ballykelly, 20

ⓔ 3/3 marks awarded   All the values are correct, for 1 mark each.

(b) Complete the nearest neighbour analysis (R_n calculation) and state the type of distribution shown in Figure 13. The nearest neighbour index equation and significance graphs are shown on pages 19 and 22, respectively. Comment on what this result indicates about the hypothesis stated.   (4 marks)

(b)

$$R_n = 2d\left(\sqrt{\frac{n}{A}}\right)$$

$$R_n = 2 \times 17.2 \times \sqrt{\frac{20}{22,275}}$$

$$R_n = 34.4 \times \sqrt{0.0008}$$

$$R_n = 34.4 \times 0.028$$

$$R_n = 0.96$$

The distribution is significantly random.
However, this hypothesis would have to be rejected as random is at the 95% level — not enough points have been used in this case.

ⓔ 6/6 marks awarded   The formula has been used properly and the type of distribution is correctly identified. The link to the hypothesis is also correct. This is a well worked out answer.

ⓔ There are 4 marks available for the successful application of the whole formula and 2 marks for the comment on the hypothesis. The distribution of the weather stations is nearly perfectly random, with a slight tendency towards clustering. However, as only 20 sites were measured, this random result is considered to be at the 95% significance level, which means that we should reject the hypothesis because we do not have enough information to say that this is significant. 1 mark is for noting the random pattern and 1 mark for suggesting this this means that the hypothesis would have to be rejected in this case.

(c) When using nearest neighbour analysis to identify a distribution pattern, a number of different factors can influence the final result of the R_n value. Describe and explain one factor that could affect R_n values.   (3 marks)

(c) This R_n value was greatly affected by the fact that there were not enough points available for the actual statistic to be used properly. Ideally, 30 sites are needed for the statistic to work properly. If 30 sites were used, this might show that places are quite regular (and the R_n will increase) so that the whole of NI is covered.

🄔 **3/3 marks awarded** This answer includes some useful points and argues clearly how the R_n result might actually change, making a valid statement that describes and explains how the R_n could be affected.

🄔 **Most answers will involve a comment that discusses the influence of area on the R_n value. They should be able to effectively describe how a larger area will usually produce a more clustered result. However, in this case it is likely that they make reference to the number of points as well — as the points were randomly scattered and this does mean that a random distribution was a distinct possibility. Perhaps the area chosen for the map was too big in this particular case.**

(d) **Study the table below. It shows the results of an investigation that measured the depth of the river at 30 cm intervals across a meander section of the Glenarm River in County Antrim.**

Site	Distance from the inside bank/cm	Depth of river/cm
1	0	0
2	30	5
3	60	7
4	90	14
5	120	23
6	150	29
7	180	35
8	210	45
9	240	45
10	270	33
11	300	50
12	330	58
13	360	67
14	390	55
15	420	43

Using the data above, plot the information on a graph and label it fully. (7 marks)

(d)

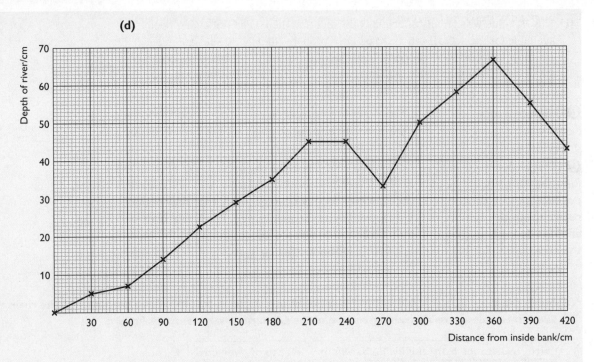

Line graph to show the depth of river and the distance from the inside of the bend

(e) **7/7 marks awarded** The title is clear, the axes are clearly labelled and all sites are clearly and accurately marked on the graph.

(e) **1 mark would be awarded for 1–3 sites labelled accurately, 2 marks for 4–8 sites labelled accurately, 3 marks for 9–13 sites labelled accurately, 4 marks for all sites labelled accurately and marked. 1 mark is for the title, 1 mark for labels on the x- and y-axes and 1 mark for making sure that line connections are accurate. Either a correlation or a line graph is acceptable.**

(e) Describe the pattern on your graph from (d). (3 marks)

(e) The graph shows that on the inside bend the depths are quite shallow (0 cm at the start and then 30 cm across the river the depth is only 5 cm). As you move across the river from one side to the other, the depth increases, though there is a patch in the middle of the river where the riverbed flattens out and there might be the odd bigger stone sitting. The deepest part of the river (at 360 cm across the river) is 67 cm deep, and the river starts to get shallower towards the far side.

(e) **3/3 marks awarded** This answer includes some good detail and uses figures to show the main features and changes in the depth of the river. Good understanding is shown of both sides and aspects of the river meander.

ⓔ **You need to describe the relationship between the two variables on the graph. There must be some clear description to show that the depths increase towards the other side of the river. Figures must be used in order to score the maximum 3 marks.**

(f) Suggest *one* geographical reason to explain this pattern.   (3 marks)

(f) The river gets deeper as you move from one side to the other because of the process of erosion. Towards the outside of the river bend the deepest part of the river is also the place where the water moves the fastest and erosion here (in particular hydraulic action) will be strong.

ⓔ **2/3 marks awarded**   This is a good answer that deals with the erosion aspects and how the deep sections of the river are formed, but it ignores the shallow sections where there might be evidence of deposition.

ⓔ **The most common answers here will address the differences in velocity and discharge on the two sides of the river, and the link with energy and erosion. You need to make sure that you refer to the features on both sides of the river to get 3 marks.**

(g) A group of geography students is asked to complete a questionnaire survey to investigate where the parents of the school's pupils were born for each year group. The school has seven year groups, with 100 people in each year group.
   Identify and explain which sampling method you think might be the most appropriate for the geographers to use to carry out this questionnaire.   (3 marks)

(g) You would need to get an equal sample of responses from each of the year groups, so you would organise this so that you got the same number of questionnaires from each of the year groups in the school.

ⓔ **1/3 marks awarded**   The student has not identified the type of sampling but then does go on to describe accurately how this might be carried out, although even this needs to be developed further.

ⓔ **The main type of sampling that would be appropriate here is stratified sampling. The researchers would need to get an equal sample/number of responses from each of the year groups. Random sampling would not work because this might group the responses into one year group more than another.**

(h) Suggest how many questionnaires you think the geographers might need to sample this particular case, and briefly explain your answer.   (2 marks)

(h) To get an accurate example of the information, each year group could get 50 questionnaires — 25 for boys and 25 for girls. This would be 50 × 7, so a total of 350 surveys would be taken.

ⓔ **2/2 marks awarded** This answer addresses the question well and explains how many surveys might be taken and how this might be broken down.

ⓔ **There is no definitive number of questionnaires that might be taken in a survey such as this — it depends on how accurate and representative the researcher wants to be. The 2 marks are for a well-argued and sensible approach. Generally, no less than 10% should be taken but it might be better to take a representative sample of 20% in this particular case. This means that between 10 and 20 questionnaires should be taken from each year group (with half going to boys and half to girls). This means that a minimum of 70 questionnaires might be needed.**

Question 2 **Population short questions**

(a) Describe *one* problem with population data collection in an **LEDC** and *one* problem associated with population data collection in an **MEDC**.

(4 marks)

Student answer to question 2

(a) In LEDCs a lot more people in the country cannot read and write and this makes it difficult to collect accurate information. They cannot read the questions themselves and need more help, and they might not even be aware of when they were born.

In MEDCs the biggest problem is making sure that information is filled in accurately. For example, some students might live at university but they might be counted in a house in the city or at their parents' home. Some people will be counted twice while others will deliberately not put their name on the form in case they have to pay more tax.

ⓔ **4/4 marks awarded** Not all the information in the answer is totally relevant but there is some good depth on the problems in LEDCs, and some balance in looking at issues of accuracy in MEDCs.

ⓔ **A range of problems could be discussed here. The answer needs to have balance, describing one problem in LEDCs and one in MEDCs. There is no need for comparison between the two problems. Answers might include the size of countries, language and access problems, gender issues, funding and training problems and identity fraud.**

(b) State what features of a population pyramid might lead you to consider that the population was experiencing a youthful population structure.

(2 marks)

(b) A youthful population usually occurs in poor countries. The base of the pyramid is wide because birth rates remain high. Death rates can also be high because people do not live long and die before they reach 60.

ⓔ **2/2 marks awarded** This discusses two elements and includes some reflection on the shape of the pyramid.

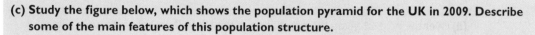 **A youthful population structure is usually found in LEDCs. It has a wide base, which indicates a high birth rate (above 30 per 1,000). The death rate is also relatively high and means that the sides are tapered sharply because people do not live long. Life expectancy can be low (but has been increasing in recent years) and few people live beyond the age of 65. Answers should describe at least two features (from BR, DR and LE) for 2 marks.**

(c) Study the figure below, which shows the population pyramid for the UK in 2009. Describe some of the main features of this population structure. (3 marks)

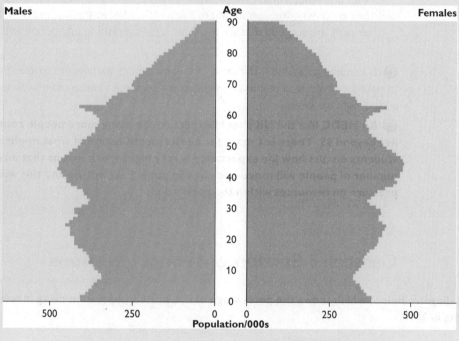

Population pyramid for the UK, 2009

(c) The population structure of the UK can be divided into three parts. Between ages 0 and 15 we can see that there is a slight increase in the birth rate (especially in the 0–5 age group). From 15–60 the sides are quite straight, which shows that the death rate is low. Most people who live to the age of 15 will reach the age of 60+. Finally, there is a sizeable number of people living beyond the age of 65, which shows that life expectancy is increasing.

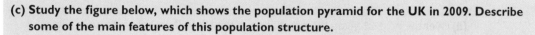 **3/3 marks awarded** This answer is broken down into three sections and then attempts to explain the three different sections of the pyramid.

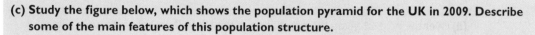 **The population structure of the UK in 2009 is fairly typical of an MEDC. Birth rate fluctuates a little, with more births having taken place in the last 5 years than in the last 20 — so birth rate was probably a little higher than usual. The sides of the pyramid are fairly straight, which indicates a low death rate, though there are two small peaks (at 25 and 43) caused either by a baby boom or as a result of in-migration.**

A sizeable number of people live beyond 65, which shows that the population in the country is gradually ageing. There needs to be good description that covers at least two of the three main components of the pyramids (BR, DR and LE).

(d) Identify and explain *one* aspect in which you might expect this population to change over the next 25 years.

(3 marks)

> **(d)** Over the next 25 years the population of the UK will probably continue to get older and older. All the people from 40 to 65 will continue to survive and will join the people who are continuing to live to 90+. The reason for this is that health services are so good that we have cures for everything and people will survive.

🅮 **2/3 marks awarded** This makes some good points but needs to develop the explanation to make sure that it is as clear as possible. Explanation needs to focus more on the reasons why people might live longer.

🅮 **An MEDC like the UK should expect to see many more people continuing to live beyond 65. There is 1 mark for a brief identification of what might change. Most students discuss how life expectancy is very high, which means that an increasing number of people will move into the 65+ zone. Few will die and this will put pressure on resources within the population.**

Question 3 Settlement short questions

(a) Study the diagram below, which shows the percentage of the world's population living in urban settlements in 2010.

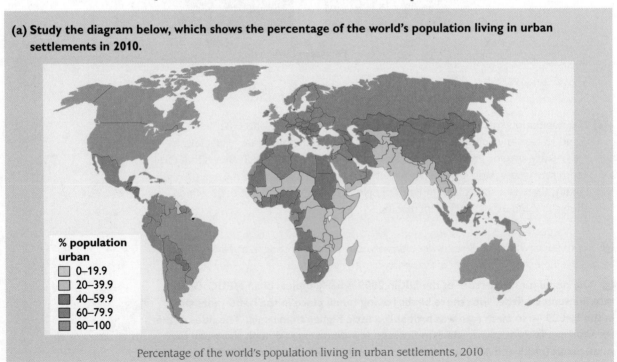

% population urban
- ☐ 0–19.9
- ☐ 20–39.9
- ☐ 40–59.9
- ☐ 60–79.9
- ☐ 80–100

Percentage of the world's population living in urban settlements, 2010

Describe and explain the distribution of areas around the world that have high percentages of urban and rural populations.

(6 marks)

Student answer to question 3

(a) From the map I can see that in places like USA, Canada, Brazil and Saudi Arabia there is a high percentage of people (80–100%) who live in cities. There are also quite a lot of people in the western European countries (60–79.9%). In contrast to this, many of the countries across Africa and Asia have a low percentage of people living in urban areas. Many of the East African countries and India and Pakistan have only 20–39.9% of people living in cities. The reason for this is that the countries with many people living in cities seem to be the richer/MEDC countries. People want to live in a city as this is where all the jobs are, and people know that living in a city will give them better lives.

ⓔ **5/6 marks awarded** This begins with a good description of the distribution of urban and rural areas, so 3 marks are awarded. However, the explanation is a little weaker and needs more depth for full marks.

ⓔ **A description of the pattern of the distribution is required before attempting to explain this pattern. 3 marks are for description. The places that have a high percentage of urban population (80–100%) include the USA, Canada, Argentina, Brazil and Australia. Credit can be given for locations using continental titles. The places with a low percentage of urban population are generally found in the 20–39.9% category and are in Africa and Asia. This is where people still largely live in rural areas. The other 3 marks are given for an explanation of how most of the urban areas are rich/MEDC countries and the rural areas are found in many of the poorer/LEDC countries.**

(b) **With reference to your case study, discuss the social and economic deprivation in an inner city area of an MEDC.**

(6 marks)

(b) In Belfast there is a lot of social and economic deprivation in the inner city. Parts of east Belfast, in Sydenham and the Newtownards Road, have been suffering for years because of the decline of the manufacturing industry in this area. This has led to massive amounts of unemployment in the area — currently 14% of the people in the area do not have jobs (compared with the NI average of 6%). This shows that this area is hit more during the recession. Many workers do not have a good education, which makes it difficult for them to get a better job.

In terms of social deprivation, this area of Belfast has one of the highest percentages of free school meals (21%), which shows that the government needs to support the families and make sure that they get good food. Few students leave school with more than five GCSEs (about 24%).

ⓔ **5/6 marks awarded** This is a good discussion of the key ideas about social and economic deprivation in the city. The discussion about economic deprivation is probably better developed than that about social deprivation.

e This question is looking for discussion of the social and economic deprivation in your **MEDC** case study city. Any answer that leaves out either social or economic deprivation will be limited to 3 marks. There should be a discussion of the different levels of deprivation, with some mention of both areas and specific locations in the city, and reference to a range of indicators.

Level 3 (5–6 marks): A good all-round discussion of the range of social and economic indicators, and figures used to develop the discussion of deprivation in a particular case study. Specific place names are mentioned in the case study.

Level 2 (3–4 marks): Some good content but in less depth. Figures linked to indicators might be missing. There might be inaccuracies or omission of discussion of either social or economic factors. There might not be a comprehensive discussion of specific places.

Level 1 (1–2 marks): A poor answer showing a limited understanding of the question.

Question 4 Development short questions

(a) Distinguish between globalisation and trade. (3 marks)

Student answer to question 4

(a) Globalisation is the process whereby countries have been able to develop their links so that they are effectively closer to each other. The transport links and communications between countries have brought them closer together, so that some companies like Nike and other NMCs have offices all over the world. Trade is the movement of goods from one place to another as imports and exports.

e **3/3 marks awarded** This is a good explanation of globalisation, which goes on to give a definition of trade, showing the differences between the two concepts.

e **Globalisation is the process whereby the world is becoming more interconnected and interdependent. It is caused by the movement of money and capital from one place to another. Trade is the flow of goods and services between producers and consumers around the world. There is overlap between the two concepts but discussion here needs to highlight the difference. If one aspect only is discussed, no marks can be awarded.**

(b) Identify and describe *one* disadvantage of aid. (3 marks)

(b) Aid can sometimes be a bad thing for a country as it encourages corruption. When money is given to a country for a project, sometimes the funds can be redirected into the bank accounts of government officials. The money does not reach the people who really need it.

🅔 **3/3 marks awarded**   This identifies one valid issue and goes on to describe this in some depth, noting how corruption means that people might not get to see any of the aid given.

🅔 **Aid is the process whereby one country or organisation gives resources to another country. Usually aid is a positive, but the answer needs to focus on a brief discussion of one disadvantage. 1 mark is for identifying a disadvantage and 2 further marks are for describing this in depth.**

(c) With reference to your national case study, discuss the pattern of regional variations in development.

(6 marks)

(c) In Italy there is an obvious difference between the richer industrial north and the poorer agricultural south, or the Mezzogiorno region. In the north, the area is flat and fertile, which means that farming in places like Tuscany is successful (wine and olives). However, the north is dominated by industry — there are many industrial cities like Turin and Milan, which have grown up to provide goods for the European market. This contrasts with the much poorer south. In the south, the soils are poor and the weather conditions are usually hot and dry. Farming is subsistence farming and this has led to many of the young people moving away from the area.

🅔 **5/6 marks awarded**   The answer goes into some good depth to describe the variation between the north and south. There is some discussion of the variation and the places, though a few more facts and figures would help to develop the answer further.

🅔 **Reference to any relevant national case study is acceptable. Answers need to describe the regional contrasts in development related to the case study. Any discussion about regional variation should contain specific place detail.**

Level 3 (5–6 marks): Contrasts in the level of development are clearly described using facts, places and figures to support.

Level 2 (3–4 marks): A good answer which is accurate but there is less factual detail, less locational information or no use of figures.

Level 1 (1–2 marks): Answer lacks depth or detail.

Question 5 **Population essay**

'The distribution of population in a country is linked to the availability of human and physical resources.'

Discuss this statement with reference to your case study at the national level. (12 marks)

Level 3 (9–12 marks): A good, detailed description of the distribution of the population in one country, with specific reference to figures and places. The answer shows clear understanding of resources and their link with the distribution within the case study. Both physical and human resources are fully addressed.

Level 2 (5–8 marks): An accurate description of the distribution patterns but with less factual detail and depth in the answer. Some answers might discuss only physical or human factors.

Level 1 (1–4 marks): A limited answer that lacks depth. There might be some basic description but the answer lacks explanation and detail. The case study might be inappropriate or incorrect.

Student answer to question 5

In France the population is unevenly distributed. Half of the total population is located on 50% of the land, mainly around the Paris basin area and the Mediterranean coastline. There are many physical factors that lead to this uneven spread of population. Firstly, around the Paris area the land is low lying and fertile, with many river tributaries meeting just outside Paris. This fertility allows for crop farming and so there is a plentiful food supply. Also, the low lying land allows for building to occur easily. The climate consists of a hot summer and mild winter, with average rainfall for France. These factors all contribute to a dense population in this area.

In the Massif Central and Armorican Massif in central France the population is sparsely populated. This area is hilly with poor, infertile soils. No crop farming can occur here and this means it is left to hill sheep farming. This leads to large areas of the land being free from any houses or communities. The climate is mainly cold and wet, with above average rainfall and cold, wet winters. It is a combination of these factors that leads to a sparse population. In the Alps region it is mainly mountainous with large amounts of snow during the winter. This makes any type of farming difficult. Also during the winter many areas become inaccessible. There is often difficulty when building and these factors lead to many areas being sparsely populated.

There are also human factors that influence population distribution. In France, urbanisation didn't occur until the 1930s. There was a sudden population increase into the cities and in particular Paris. The suburbs of Paris quickly established themselves and they are now heavily populated. Also with such a large collection of people and industry in one area there is a large population threshold. Therefore any surrounding industry struggles to compete and this results in a large area around Paris unable to support any industry.

There are also political factors. The French government decided to relocate many of its offices to different areas around France in order to try to repopulate other areas and depopulate Paris. For example, the tourism board is located in southern France and the education board in eastern France.

● **8/12 marks awarded** This is an upper Level-2 answer. There is some discussion of the variations in population distribution in France but the answer is limited and needs more discussion and detail in relation to both the physical and human factors.

The answer needs to make more specific reference to places and figures to support the answer — for example, the total population of France is 64 million, with 10 million or 15% of the population living in rural areas. The population density is around 117 people per km^2. More reference to place would have been good — as well as the Massif Central, areas such as the southern Alps, the Pyrenees, Corsica, Limousin, Auvergne and Aquitaine are sparsely populated. The Ile-de-France (Paris), Rhône-Alpes, Provence, Toulouse, Brittany and Nord-Pas-de-Calais are more densely populated.

The discussion of different physical and human factors could have more depth — for example, when talking about the Massif Central, reference could be made to high mountains (peaking at 1,886m). These mountains have a wet, cold climate in the winter, which often leads to snow. Soils can be thin and cannot support vegetation. Grass will not grow, which means that it is not suitable for either pastoral or arable farming. There are few energy resources and people find it a hard place to live. A good answer needs depth, facts and figures when referring to both physical and human resources.

A good way of organising this answer might be to use the planning box technique. Students could structure their answer/plan in the following way and allocate time accordingly.

Planning box

'The distribution of population in a country is linked to the availability of human and physical resources.' Discuss this statement with reference to your case study at the national level. (12 marks)	
Population distribution: sparsely populated (mostly due to physical factors)	Population distribution: densely populated (mostly due to human factors)
Physical resources: climate (3 min)	Human resources: farming (links to climate and soils) (3 min)
Physical resources: relief (altitude) and soils (3 min)	Human resources: industry, transport and communications (3 min)

Question 6 **Settlement essay**

With reference to your case study concerning a protected area at a local or regional scale, describe how it has been managed for conservation, recreation and tourism. (12 marks)

● Answers need to name a protected area (e.g. the Peak District National Park) and then outline some of the specific examples of how it has been managed. There must be some mention of all three areas: conservation, recreation and tourism. Places names and figures are expected for a Level-3 answer.

Level 3 (9–12 marks): Solid reference to a named protected area and how the area has been managed for conservation, recreation and tourism. Good depth in relation to all three. Specific reference to the case study material and quality of communication is relevant and good.

Level 2 (5–8 marks): Answer is still good but the depth of knowledge in relation to the case study may be less. Answer has omitted one of the management issues (recreation, tourism or conservation). An answer with poor understanding of the case study might be limited to this level.

Level 1 (1–4 marks): Answer lacks depth in relation to the case study or might only deal with one of the management issues adequately. Quality of written communication is poor and/or limited.

Student answer to question 6

The Peak District became the first National Park in 1951 and is now the fifth largest National Park in the UK. It is situated in the north of England and covers 555 square miles.

National Parks can be defined as conserving and enhancing natural beauty, wildlife and cultural heritage, and promoting opportunities for the understanding and enjoyment of their special qualities.

National Parks were set up with the specific purpose of protecting areas of natural beauty in the countryside. Conservation in the PDNP involves more than just preventing damage and leaving things alone. It involves maintaining the best features of the landscape, for example listed buildings including Haddon Hall, and improving neglected features such as stone walls and replanting old woodlands. Also managing development so that damage is limited. This involves restrictions on new buildings and recreation activities such as certain water sports.

Conservation of villages is also important, with the PDNP authority controlling the erection of new buildings, limiting the range of building materials and creating conservation areas in many villages that have buildings of historic or architectural interest.

In 1999 the PDNP authority issued a management plan to set out a vision for the future. This vision is based on three main aims. Firstly to conserve and enhance the special qualities of the National Park, secondly to provide opportunities for visitors' enjoyment and understanding, and finally to improve the quality of life for people who live in, work in and visit the area.

Recreation is the activity of leisure over a short period of time, for example for 1 day. The PDNP management plan has begun to manage recreation in many ways. Firstly a cycle hire scheme and special cycle routes along disused railway lines or quiet roads have been set up. Fishing and sailing are permitted only in a small number of the many reservoirs and motor sports are strictly controlled to avoid damage to the environment, through noise, air and water pollution.

Traffic calming measures have also been put in place across the most scenic routes, including the south Pennines, to deal with traffic problems. Also, the use of public transport is promoted to help relieve congestion and to allow ease of travel, for visitors and also those who live and work in the park, during peak times.

Since its designation as a National Park the variety of tourist activities has increased. Just over 30 million day visits occur each year. It is argued by some that tourism is by far the most important source of income in the Peak District National Park, providing around 500 full-time jobs, 350 part-time jobs and 100 seasonal jobs. Tourism can provide the income to keep many historic buildings in good repair. Local people also benefit by providing caravan and camping sites in their fields and offering bed-and-breakfast accommodation in their homes. Local shops also benefit from this tourist trade. The popularity of honeypot villages such as Castleton means there is a greater level of local employment than is usual in a village of its size.

In its vision for the future the PDNP authority states that it will 'provide opportunities for visitors' enjoyment and understanding'. It has done this in many ways, including in 2000 holding over 300 local and traditional events, reflecting the importance of local history and customs. It is also increasing the number or capacity of car parks in popular villages and at beauty spots, often with public toilets and information boards. Picnic areas are often close to car parks. Facilities like these are carefully designed to minimise their impact on the landscape.

11/12 marks awarded   This high Level-3 answer is comprehensive and shows a command of the case study as well as application of knowledge to the question. It discusses how each of the management issues, in relation to conservation, recreation and tourism, affects the Peak District National Park. There is good balance in the answer.

There is some good discussion of the PDNP and reference to the management plan for the National Park. However, the answer does tail off a bit towards the end, where content about the management of tourism could have been taken further and talked about the slight differences between how tourism and recreation are managed.

Planning box

With reference to your case study concerning a protected area at a local or regional scale, describe how it has been managed for conservation, recreation and tourism.   (12 marks)
Brief introduction to case study of protected area (e.g. Peak District National Park)
How has the PDNP been managed for conservation? (4 min) Soil erosion Damage to wildlife and farmland Protecting biodiversity, ecosystems and heritage
How has the PDNP been managed for recreation? (4 min) Congestion of villages and beauty spots Managing recreational activities
How has the PDNP been managed for tourism? (4 min) Managing the impact on local services Managing the pressure on services

Question 7 Development essay

Globalisation and trade have had a big impact on the development of LEDCs. Define these issues and show how they have affected the development of LEDCs in a positive and/or negative way.   (12 marks)

Globalisation and trade are both issues that can have a big impact on the development of countries. The answer requires a definition of each of the issues and then a deeper explanation of how the process has played a role in affecting the development of LEDCs. A maximum of Level 2 will be awarded if no definition of each process is given.

Level 3 (9–12 marks): A detailed account that shows a solid understanding of the main components of both processes. Clear understanding is shown of the positive and/or negative effects that each process brings to LEDCs or to particular named LEDCs. Good use of terms to support the answer.

Level 2 (5–8 marks): Some good depth in the answer but level of detail might be less. Students who leave out one process completely will be limited to this level.

Level 1 (1–4 marks): Answer might lack knowledge of the issues under discussion. Answers that focus only on the definition but have no elaboration of the effect on development will be limited to this level.

Student answer to question 7

Globalisation has had a big impact on the development of LEDCs. It is the process where the different countries become more connected and dependent on each other. This is mostly caused by the increase in trade and uses much of the new technology and communications to ensure that products can be moved from one part of the world to another.

Globalisation can have a big impact on countries as this often brings much needed hard cash into poor LEDCs. Many multinational companies set up factories and offices all over the world — they want to be able to make products cheap and then transport them for sale in the MEDCs. Globalisation brings jobs to people in the LEDCs and this provides formal jobs, training and a stable source of income. This money will then filter through the whole economy. Sometimes the MNCs will also spend money helping the LEDC to improve its transport infrastructure — there will be investment in ports and motorways. Workers will be able to get new skills and will become better educated.

Trade is when goods and services are flowing from one country to another. Anything that comes into a country to be sold is called an import and anything which is made in a country and sent to another country for selling is called an export.

Trade is an important way that countries can develop and get richer. Many poor countries can make money through sending primary products like food, oil and natural gas to the MEDCs. The countries will make money on these products but they would have made more money if they were allowed to process the raw material before it was sent somewhere else. In Ghana, much of the wealth that the country has was built on the trade of cocoa, gold, diamonds and timber. These all bring much needed money into the country, which can then be spent on healthcare and education and in trying to develop the lives of the people who live there.

ⓔ **8/12 marks awarded** This does include definitions of the two different concepts — globalisation and trade. However, the explanation of how these two ideas have affected the development of LEDCs could have been taken further. The discussion on globalisation is good but lacks specific details about a country or countries. The discussion of trade is a lot less detailed and although it does talk through how trade might have affected the country of Ghana, this could be explained in a bit more detail.

Planning box

Globalisation and trade have had a big impact on the development of LEDCs. Define these issues and show how they have affected the development of LEDCs in a positive and/or negative way. (12 marks)	
Globalisation	*Trade*
Definition: what is globalisation? (2 min)	Definition: what is trade? (2 min)
How has globalisation affected development in LEDCs? (4 min) Positive impact Possibly negative impact	How has trade affected development in LEDCs? (4 min) Positive impact Possibly negative impact

Knowledge check answers

1

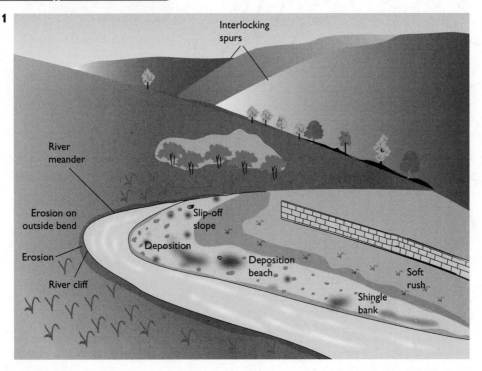

2 (a) Section D
  (b) Eagley Way and Darwen Road

3 (a)

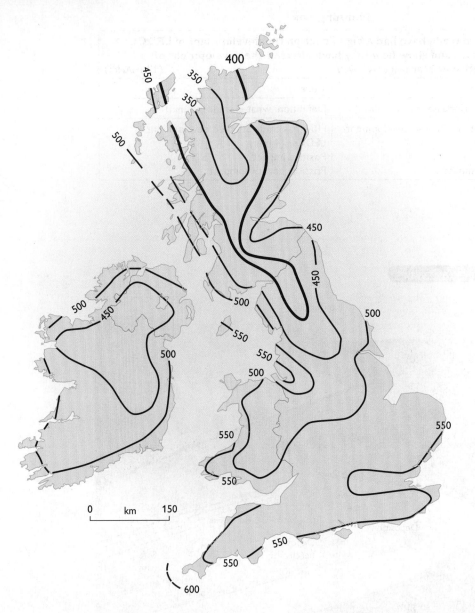

(b) The lowest amount of evapotranspiration is found towards the north of Scotland (less than 350 mm). The amount of evapotranspiration increases moving south, with high amounts found along the Welsh coast in the west and along the south coast of England (550 mm). The maximum is found on the tip of Cornwall in the southwest of the British Isles.

4 The graph is a bar chart.

It shows that 45 of the people surveyed came from Coleraine. This was the largest category, followed by Portrush (17), Ballymoney (13), Bushmills (11) and then Ballymena (10), Portstewart (9), Derry (5) and Ballycastle least with 4. The survey was conducted in Coleraine so the majority of the people surveyed would be local. Generally, the further away you travel from Coleraine, the fewer people came from that place.

5 (a) 25% hydroelectric, 52% nuclear, 23% thermal and other

(b) 10% hydroelectric, 48% nuclear, 42% thermal and other

6 (a) Mean is 74, median is 74, mode is 76

(b) The mean because it is the only measure that makes use of all of the values.

(c) The mode because it only records the value that occurs most times and this could be found at either extreme.

7 (a) This value is 99.9% significant.

(b) The Spearman's rank correlation shows that there is a strong negative relationship between the two variables.

The reason for this is that in many countries it would be expected that, as the GNI PPP per capita (amount of wealth in the country) increases, so the number of births in the country would decrease. This might be due to a number of factors such as better access to contraceptive methods, better education and career prospects for women.

8 (a) Usually the minimum number of distances that needs to be measured is 30. In this case there are only 20 places/distances and this might call into question the validity and accuracy of the result.

(b) If the map was focusing on one area, the R_n value would increase and the pattern would be more regular.

9 The census helps government departments to plan for the next 10 years, looking at key trends so that decisions can be made in relation to the size of population, education, health and disability, housing, employment, ethnic groups and transport.

10 The main issues of census reliability in MEDCs are usually down to human error. Mistakes can be made in making sure that all areas are covered and that accurate maps/housing areas are drawn up. Some people are wary about the use the government will make with the information and they may supply incorrect information. Diverse language needs and special needs might also mean that coverage is not universal.

11 Reliability issues for vital registration in an MEDC might be that people do not follow the correct procedures to get their children registered, or they might decide they do not want their children to be registered. Although the system is relatively simple it can be easily avoided.

12 The young dependent population (aged between 0 and 15) can increase (in LEDCs) or decrease (in MEDCs). The economically active population (aged 16 – 64) will usually increase in both MEDCs and LEDCs. However, the sides of the pyramid will usually be straighter in MEDCs because the life expectancy increases, showing that most of the people who turn 16 will reach the age of 64. The aged dependent population (aged 65 and beyond) can increase massively in MEDCs and will increase at a much slower rate in LEDCs.

13 Between 1750 and 1800, the birth rate was high (above 37 per 1,000) and the death rate was high (31 per 1,000). The population was shaped like a concave pyramid with a wide base and the sides tapered in towards the top. Life expectancy was low and few people reached the age of 65.

Between 1800 and 1880, the sides of the population pyramid expanded outwards as life expectancy increased from 45 to 55. People were living longer so the death rate fell to 19 per 1,000. The birth rate remained high (at 30 per 1,000) and the base of the pyramid remained wide.

Between 1880 and 1950, life expectancy continued to improve and more people lived to the age of 65. Death rate fell to a very low level (13 per 1,000) but the main difference in the population shape was that the birth rate had started to decrease (down to 16 per 1,000).

From 1950 to the present, the population of the UK has stabilised. Life expectancy continues to improve (up to 75) and the death rate is very low (9 per 1,000). The birth rate is now very low (13 per 1,000) but has increased slightly due to immigration in recent years.

14 The Xinjiang area of China is rural. There are slightly more births here than in other parts of China. Many people are required to work the land but others struggle to survive and life expectancy is low. In Guangdong, the population is more urban and birth rates are low as people concentrate on careers rather than starting families. There is a large young working population but this is mostly a direct result of migration.

15 The dependency ratio is calculated as the total of the youthful dependent population (aged 0–14) and the aged dependent population (65+) divided by the working population (aged 16–64). A dependency ratio of 50–75 indicates an MEDC; in LEDCs the ratio is closer to 100.

16 Economic: as people live longer owing to improvements in healthcare, they require pension payments for longer periods of time. In the past, people retired at 60 and died at 70 but now retirement at 65 and death at 85 is usual, which means 10 extra years of pension payments.
Social: families have to take more responsibility for the care of their elderly relatives as they start to suffer from degenerative diseases.
Political: people are increasingly concerned about decisions they can take in relation to their right to die when they want to. Governments have to consider policies to deal with euthanasia issues.

17 Economic: with so many young people in the population it is difficult for the government to pay for a quality education system for all. The education authorities have little money to invest and parents have to pay for their children's education.
Social: often people die from preventable diseases. Life expectancy is low, death rates are relatively high and people die from illnesses that are treatable in other parts of the world. Families can struggle to access quality healthcare.
Political: governments in LEDCs often do not consider the needs of young people as a priority — they think that any extra money should be spent on icons of wealth and power rather than on education, healthcare or welfare.

18 Greenbelts are a planning mechanism used to control and to stop the building of houses on greenfield sites.

19 Suburbanisation is when people living in the city (usually the inner city) move out into the suburban areas of the city. Counterurbanisation is when people move from the inner city and the suburbs to places further out of the city — towards the surrounding rural towns and into the rural–urban fringe.

20 Most of the impacts are not positive — the sprawl brings traffic congestion and pollution. Because more people want to move into rural housing, competition for land and houses increases and house prices rise. Green space and animal habitats come under increased threat. On the positive side, the rural–urban fringes might start to develop and their services may improve.

21 The movement of people out of the inner city frees up valuable land space, which can then be redeveloped. People can then

build better homes and houses than they might have had before. However, as people move out, jobs, shops and services might also move out and the city might suffer as a result.

22 When decline in an area starts, the young people are often the first to leave — for a better education and to extend their horizons for better jobs, services and entertainment/lifestyles.

23 The Peak District National Park area is used for a variety of activities that can cause damage and erosion of the landscape. Wildlife habitats and heritage can be endangered. Management strategies are needed to work with farmers and landowners to protect the land but to ensure that it is used and not abused. Fragile environments and areas of biodiversity are protected.

24 People come into the Peak District National Park to take part in sport, leisure and recreational activities. This can create problems of noise pollution and erode the landscape. It can also create congestion issues in villages and beauty spots. Plans need to be drawn up to manage the placement of these activities, and control measures need to be put into place to manage the traffic and ensure that congestion does not build up.

25 People come to stay in the National Park area for short periods of time, mainly during the summer when the weather is better. This concentrates the number of visitors who are likely to be in an area at any one time, which can put a strain on services and cause the land use to change to cater for the tourist economy. Sustainable planning policies to manage tourism and the impact it will have on the local community are put into place to ensure that local communities benefit from tourism but do not feel threatened.

26 Redevelopment is when an area is demolished and redesigned. In the inner city this might mean that a street of terraced houses is knocked down and replaced with a block of flats or other cheap housing. Gentrification is when an area is demolished and redesigned — but the original residents will not move back. Instead, richer people might be attracted to live in the new, expensive accommodation.

27 People usually live in poor, temporary accommodation. The slum areas are often on poor quality land — low-lying areas at serious risk of flooding or poor soils, close to rubbish tips or steep slopes where landslides are common. The houses are made of any recycled or reclaimed materials that people can get their hands on. The houses are small and lack even the most basic services. Houses are often shared with one other group or family and 'hot-bedding' is common.

28 Gross national product (per capita) is the total economic value of all the goods and services provided in a country through the year (including output generated in other countries), divided by the number of people who live in the country. GDP (per capita) is the market value of all the goods and services provided within a country through the year, divided by the number of people who live in the country. It is usually seen as a measure of the standard of living of people in a country.

29 Car ownership, which shows the number of people who have enough disposable income to buy personal vehicles.

30 The age that people live to is often an indicator of the health and social care within a country. If people lack basic medical services they will not be expected to live long but where the medical facilities are good, people will live longer.

31 Life expectancy, mean years of schooling and GNI per capita ($).

32 The PQLI measures infant mortality rate, life expectancy (which both indicate the how good the healthcare is) and literacy rate (which is a basic measure of education). These are all social measures that require huge amounts of political investment. The HDI uses one economic indicator, one social indicator and a more robust indicator of education — one that measures the access that children have to education rather than just something that can be achieved in a relatively short time.

33 The south of Italy has not developed to the same extent as the north. Traditionally, the south has been more focused on agriculture and, with its hotter, drier climate, it was much slower to move towards industrialisation. Location has a major part to play too — the north is on the doorstep of Europe while the south is much further away and on the European periphery.

34 Many of the colonial powers were good at developing the social and economic infrastructure of their conquered lands — they built government buildings, schools, hospitals, roads and railway lines, which all helped to develop the economic wealth of the area. However, many of the areas were exploited, and mineral wealth, raw materials and in some cases slaves were removed and taken to other places around the world.

35 Globalisation is the process in which the world is becoming more interconnected and interdependent. Trade arrangements, goods and services are easily moved around the world. A major indicator for globalisation is the rise of transnational companies which dominate and often control the manufacture and trade in an area.

36 Bilateral aid is when aid is given from one country to another. The aid is usually tied so that an MEDC can direct the money towards particular issues and priorities. Multilateral aid is when aid comes from world/international organisations such as WHO and the UN.

37 A trade deficit is when there are more imports coming into the country and few exports, resulting in more money leaving the country than coming in. A trade surplus is when there are more exports leaving than imports coming in, resulting in more money coming into the country than leaving it.

38 MEDCs are often saddled with having to write off the debt that has built up in LEDCs, and this can cause economic issues. MEDCs are much less likely to agree to lend further money to countries where debt has been written off.

39 Ghana is located on the main sea-based route that Europeans took to get around Africa to Asia. Over time, the location and climate meant that gold, natural resources such as gas and oil, and farm produce (e.g. cocoa beans) could be exported.

A

Africa

   Ghana case study 66–67

   Nairobi, Kenya, case
     study 54–55

aged dependency,
   implications of 34–35

ageing population, rural areas 43

aid 64, 67

annotated sketch maps 10–11

Areas of Outstanding Natural
   Beauty (AONBs) 45

B

bar graphs 11

Belfast, case study 50–52

birth rate 27

birth registrations 26

C

carrying capacity 45

census data 23–25

central tendency,
   measures of 15–16

China, population
   pyramids 31–33

choropleth maps 8–9

colonialism 61–62, 66

conservation management 45,
   46–47

counterurbanisation 42–43

crude birth/death rates 27

D

data collection 6–7, 24–25

data processing 6, 7–23

death rate 27

death registrations 26

debt 65, 67

dependency
   (neo-colonialism) 62, 66

dependency ratios 33–34

implications of 34–36

depopulation of rural areas 44

development

   composite measures 57–59

   definition problems 55

   economic measures 56

   regional contrasts 59–61

   social measures 56–57

development essay 89–91

development issues

   aid 64

   colonialism 61–62

   debt 65

   Ghana case study 66–67

   globalisation 62–63

   neo-colonialism (dependency) 62

   trade 64–65

dot distribution maps 8

E

economic activity 55

economic deprivation 51

economic measures of
   development 56

economic regeneration 48–49

embedded skills 6

essay questions

   development 89–91

   population 86–87

   settlement 87–89

examination skills 68–69

F

field sketches 10

flow line maps 8

France, population
   distribution 38–39

G

gentrification 51, 52

Ghana case study 66–67

globalisation 62–63, 66

graphical skills 11–14

Great Britain, change in
   population over time 30–31

greenfield developments 41–42

gross national income (GNI) 56

gross national product (GNP) 56

H

healthcare 56–57

Highlands and Islands
   Enterprise 48–49

human development index
   (HDI) 57–58

I

index of rurality 41

infant mortality rate 57

informal settlements 54

in-migration 27

inner city challenges 49–52

isoline maps 9–10

Italy, regional
   variations 60–61

K

Kenya

   Nairobi case study 54–55

   reliability of vital
     registration 26

L

LEDCs

   census taking in 25

   rapid urbanisation
     issues 53–55

   youthful dependency 35–36

life expectancy 56–57

line graphs 12

M

map skills 7–10

mean 15